Anaerobic Digestion Foresight: Using Models to Predict UASB Reactor Instability

Shah

TABLE OF CONTENTS

1 INTRODUCTION

1.1 BACKGROUND

Anaerobic digestion (AD) is a biological process mediated by a consortium of various organism groups, (Ahring, Sandberg & Angelidaki, 1995; Sötemann et al., 2005b) resulting in fermentation and degradation of organic material. AD is very commonly used to stabilise sewage sludge and degrade industrial organic wastes because it generates energy-rich biogas. Furthermore, AD bioprocesses have a very low biomass yield compared with aerobic processes, thereby reducing the amount of sludge requiring disposal.

Despite its numerous benefits, AD is generally considered to be an unstable and sensitive process (Graef & Andrews, 1974; Chen, Cheng & Creamer, 2008). The instability is typically caused by one of the organism groups being inhibited due to hydraulic overloading, organic overloading, pH and temperature changes or the introduction of a toxin to the system (Ahring, Sandberg & Angelidaki, 1995). Recent advances in research and the development of mathematical AD models have given impetus to and greater confidence in the application of AD. Improved AD models hold the promise of being capable of predicting AD instability before it happens and so prevent anaerobic digester failure, but this requires the model to be suitably versatile.

Upflow anaerobic sludge bed (UASB) reactors are high-rate anaerobic digesters which due to their physical properties, allow sludge accumulation at the bottom of the UASB reactor. The sludge accumulation results in separation of sludge and hydraulic retention times (SRT and HRT), and while the HRT is generally short, good performance results can be expected because the SRT is long. Since the UASB reactor is a high-rate digester, and large amounts of solids are present at the bottom of the reactor to mediate bioprocesses, intermediates may accumulate which requires the intermediate processes to be well-calibrated.

An AD model calibrated to replicate the conditions of UASB reactors and prediction of varying concentrations of intermediate AD products may be used for applications wherein the digester is close to failure conditions or at unstable operating conditions. One such a scenario is the start-up of an anaerobic digester. During this time, intermediate products accumulate until the biomass in the digester can sufficiently cope with the incoming load.

The pH often declines as a result and only recovers once the intermediates have been utilised. The AD start –up can be a difficult and costly process in practice, hence beginning to model its optimised control would promote profitability of the process and improvement of system health.

The Sulphate reduction, Autotrophic denitrification and Nitrification Integrated (SANI) process was developed to treat saline sewage which is generated in Hong Kong due to the flushing of toilets with seawater. It comprises an anaerobic sludge bed reactor (SRUSB – sulphate reducing upflow sludge bed) in which biological sulphate reduction (BSR) occurs, an anoxic reactor for autotrophic denitrification with sulphide as the electron donor and nitrate as the electron acceptor and an aerobic reactor for nitrification of ammonia and oxidation of residual sulphide. The SRUSB is typically subjected to varying incoming flows since it is an inline system. This results in varying bioprocess dynamics in the SRUSB, which also impacts the anoxic and aerobic reactors. To successfully model the SRUSB in the SANI system, a calibrated anaerobic model is essential.

1.2 PROBLEM STATEMENT

Though AD is an extremely beneficial way of stabilising sludge and treating other organics, the lack of information surrounding anaerobic digester failure has led to an incremental approach to modelling development. Intermediate bioprocesses are not well calibrated in the models, and consequently, they cannot model AD dynamics. Without adequate modelling of AD dynamics, wherein intermediate products accumulate, the AD model cannot be used for start-up, failure or even UASB reactors which may have temporary failure conditions at the bottom of the bed.

The SANI system has been documented and modelled using steady-state principles with lab-scale and pilot plant systems for validation (Lu et al., 2012a). However, knowledge is still lacking surrounding the performance under dynamic conditions. Specifically, when considering dynamic conditions, the SRUSB is expected to be particularly sensitive to the changing flows because the HRT and SRT may vary, similarly to a UASB reactor.

2

1.3 RESEARCH OBJECTIVES

This study develops a dynamic AD model which is calibrated to model a temporary failure condition at the bottom of a UASB reactor and recovery in the upper part. In the interest of applying the model to the SRUSB of the SANI system, this calibrated model would also be extended to include BSR processes using the modelling platform, WEST® (Vanhooren et al., 2003, MikebyDHI 2014). The calibrated model will be used to model a digester start-up scenario and also check the model's capability to model digester failure, thereby indicating its potential. Due to the extensive capabilities of the software, the system will be analysed under dynamic conditions, and the performance and stability of the system under these conditions will be assessed.

1.4 SCOPE AND LIMITATIONS

Due to the extensive work required for this book, this study excluded experimental work as part of the scope and therefore involved mathematical modelling only. Much detail, therefore, focussed on ensuring that the model best represents the intermediate bioprocesses using the data available from an existing dataset to calibrate it.

Although UASB reactors may have intermixing of biomass in between the layers, and possible granules which may be carried by the gas bubbles higher up the bed, this behaviour is not currently being modelled and falls outside of the scope of this investigation.

1.5 OUTLINE

The literature review begins with a discussion on AD under methanogenic conditions and the organism groups accepted to be present under such circumstances. It expands on the bioprocess rate of hydrolysis and those of the subsequent AD organism groups following hydrolysis. Significant emphasis is placed on the organism inhibition terms since these become important under conditions approaching failure. The literature review concludes with a brief description of the SANI process.

The model description provides a background to the WEST® modelling platform, with a specific focus on PWM_SA_AD which is a subset of the Plant-Wide Model South Africa (PWM_SA) developed by the Universities of Cape Town and KwaZulu-Natal. It provides details on the physicochemical framework and discusses in detail how the ionic processes are linked to the slower bioprocesses within the model.

Before calibration of PWM_SA and its AD subset, PWM_SA_AD, a detailed check on all model stoichiometry and kinetics was conducted. Following this thorough check, the stoichiometry required for the SANI process was derived and added into PWM_SA_AD. The calibration protocol that was developed is then given in detail, along with the necessary changes to the gas evolution kinetics.

As a validation check, the calibrated PWM_SA_AD model was applied to simulate anaerobic digester failure and compared with Anaerobic Digestion Model No. 1 (ADM1) under identical conditions. Digester start-up conditions with PWM_SA_AD was also investigated. Conclusions and recommendations for further AD model development are made after that.

1.6   NEW KNOWLEDGE CONTRIBUTION

To date, there have been no reported models for UASB reactors except where the UASB reactor is modelled as a completely-mixed reactor and a clear effluent layer such as that of Poinapen and Ekama (2010a). The UASB reactor model developed in this book allows UASB reactors to be modelled as a sludge bed with different bed layers wherein biomass solids accumulate in different proportions and the bioprocesses adopt their own volumetric kinetic rate dependent on the interconnected biomass and substrate concentrations.

With the model suitably calibrated to model intermediate bioprocesses, it can be used for failure prediction and start-up conditions, both of which lack modelling knowledge. Furthermore, since the HRT and SRT are separated and differ like they are in an actual UASB reactor, biodegradable soluble (BSO) and biodegradable particulate organics (BPO) can be modelled independently, and the utilisation rate of soluble readily biodegradable organics not only depends on the production from slowly biodegradable organics as in AD sludge but also on the high soluble organics concentration entering from the influent mainstream. Adequate calibration of the utilisation rates of BSO and BPO is crucial for modelling the SANI process because their degree of utilisation affects downstream processes.

In the SANI system SRUSB, its HRT needs to be sufficiently long to remove practically all the BSO therein because ingress of BSO into the subsequent autotrophic denitrification anoxic moving bed bioreactor (MBBR) adversely affects the biofilm denitrification rate. Also, any excess sulphide produced in the SRUSB not utilised for denitrification in the anoxic reactor is oxidised to sulphate in the aerobic reactor and unnecessarily uses oxygen. Because autotrophic denitrification requires significantly less electron donor than heterotrophic denitrification, particulate organics removal can be very high in the primary separation unit (PSU) before the SRUSB. This makes the SRUSB resemble a UASB reactor treating BSO, with additional supporting biomass produced from the influent BPO which remain trapped in the bed and so are subject to SRT, not HRT, except it does so sulphidogenically rather than methanogenically. This is one of the main reasons why the modelling of the UASB reactor system was selected as a starting point for modelling the SANI system SRUSB reactor.

The high particulate organics removal in the PSU before the SANI system SRUSB caters for high methane production from the primary sludge produced. The removal of solids before the SRUSB, therefore, results in a lower energy utilisation for aeration in the SANI aerobic MBBR and low sludge production from using autotrophic bioprocesses for nitrogen (N) removal, all of which are significant benefits over conventional activated sludge wastewater treatment.

2 LITERATURE REVIEW

2.1 ANAEROBIC DIGESTION

AD dates back more than a century, making it one of the oldest wastewater treatment processes (Sötemann et al., 2005b). It is a fermentation process whereby organic material is degraded, and biogas is produced (Henze et al., 2008). Furthermore, AD takes place in the absence of oxygen; thus, the sludge produced is significantly less compared with aerobic treatment. Not only can AD occur under methanogenic conditions, but also under sulphate-reducing as well as acidogenic conditions (Ristow et al., 2005).

Of the three conditions listed above, AD under methanogenic conditions is the most widely implemented and has been studied extensively. This condition produces methane and carbon dioxide as the final gaseous product (Sötemann et al., 2005a). The methane-producing capabilities of AD have resulted in its increase in popularity in recent years as a means of generating renewable energy to reduce fossil fuel usage (Chynoweth, Owens & Legrand, 2001).

2.1.1 AD Processes and Organism Groups

The organism groups required are acidogens, acetogens, acetoclastic methanogens and hydrogenotrophic methanogens (Mosey, 1983). Although the hydrolysis process of complex to simpler organics is listed as a separate process, it is mediated by the acidogens, and it is seen as an initial step preceding AD. The reactants and products utilised and formed by each organism group are summarised in Table 2, which follows a detailed discussion on each process.

2.1.1.1 Hydrolysis

The hydrolysis process has been described as an extracellular step which allows the complex material to form soluble products which can be transported across the cell membrane (Batstone et al., 2002; Sötemann et al., 2005b). Soluble substrates are required for the subsequent acidogenesis, acetogenesis, acetoclastic and hydrogenotrophic methanogenesis steps of AD. The hydrolysis of complex organics is mediated by the acidogens and precedes methanogenesis. It is a complex process which is typically considered the rate-limiting step in sewage sludge AD (Sötemann et al., 2005b).

The assumed products of hydrolysis differ widely between different authors, though this is not of significant concern in modelling since they are intermediate products and unlikely to ever accumulate significantly in the anaerobic digester under normal stable operating conditions. However, hydrolysis is generally not the rate-limiting step for AD of substrates which are sugar-rich and cellulose-poor, which are instead governed by the methanogenesis process (Li et al., 2017). Also, to model AD under failure conditions, intermediate products will undoubtedly accumulate and so require all the bioprocesses to be modelled and calibrated.

In ADM1 (Batstone et al., 2002), the complex particulate material is assumed to be composed of carbohydrates, proteins and lipids and these are hydrolysed to monosaccharides (MS), amino acids (AA) and long-chain fatty acids (LCFA) respectively (Batstone et al., 2002). This follows the approach of Gujer and Zehnder (1983) who in turn supported the work of Kaspar and Wuhrmann (1978). These authors state that 67% of the total biodegradable organics forms sugars and AA and the remaining 33% forms LCFA. Masse and Droste (2000) also assumed that complex particulate material comprises carbohydrates, proteins and lipids which form sugars, AA and LCFA; however, they do not give the fractions of each.

While useful for modelling AD, carbohydrates, lipids and proteins are not measured in wastewater treatment (WWT) practices where AD is widely applied. Chemical oxygen demand (COD), Total Kjeldahl Nitrogen (TKN), Free and Saline Ammonia (FSA), Total Phosphorus (TP), Orthophosphorus (OP), Volatile Suspended Solids (VSS) and Total Suspended Solids (TSS) are more commonly used in WWT practices. So, an organics characterisation method that uses these routine tests will be more practical. Therefore, a more convenient way of expressing the complex BPO composition which undergoes hydrolysis is in a generic form, $C_xH_yO_zN_aP_b$ as done by Sötemann et al. (2005b) in the University of Cape Town Sludge Digestion Model (UCTSDM).

This is convenient because the wastewater measurements such as COD, TKN, FSA, TP, OP, VSS and TSS tests allow each of the seven organic constituents of influent wastewater and also biomass to be fully characterised in this form where the seven organics constituents are volatile fatty acids (VFA), BSO, BPO, unbiodegradable soluble (USO) and unbiodegradable particulate (UPO) with UPO and BPO being separated into settleable and non-settleable fractions.

In UCTSDM, instead of three separate products forming during hydrolysis from the BSO and BPO, a glucose intermediate is formed. This was deemed acceptable for sludge digestion because glucose is an intermediate product in AD (see Figure 2) and Sam-Soon et al. (1990) commented that the acidogens have a doubling time of 2 hours with the result that the glucose should be utilised very quickly. Mosey (1983) also indicated that the acid-forming bacteria have a short doubling time and because the pathways of glucose are well established, it seemed logical to select glucose as the intermediate hydrolysis product.

2.1.1.2 Acidogenesis

In the case of ADM1 (Batstone et al. 2002) and other preceding models which follow a similar format, acidogenesis (also called fermentation) takes place on the MS and AA formed during hydrolysis.   The fractions of the different COD products are given in Table 1 below.

Table 1: Fractions of COD Products in ADM1 in Acidogenesis

Reactant	Hydrogen	Butyrate	Valerate	Propionate	Acetate
MS	0.19	0.13	n/a	0.27	0.40
AA	0.06	0.26	0.23	0.05	0.41

Batstone et al. (2002) excluded the effect of hydrogen partial pressure on the acidogenesis of MS since the supported this effect ((Costello, Greenfield & Lee, 1991; Romli et al., 1995; Skiadas, Gavala & Lyberatos, 2000)), used equations which are not typically observed in cultures, such as the formation of propionate only from glucose.

The LCFA which are formed during hydrolysis does not undergo acidogenesis but rather acetogenesis, which is discussed below. The main difference as described by Batstone et al. (2002) is the use of an internal electron acceptor in the case of acidogenesis while for acetogenesis, an external electron acceptor is required. Acidogenesis reactions have a high energy yield and are rapid, and the acidogens can operate over a wide pH range (Angelidaki et al., 2011).

Because the UCTSDM model derived by Sötemann et al. (2005b) forms glucose as the product of hydrolysis, glucose is the reactant for acidogenesis. The acidogenesis process is assumed to be influenced by high hydrogen partial pressure. Mosey (1983) indicated that higher (than acetate) VFA such as butyrate and propionate are formed only as a result of a surge load and is not usually a product of acidogenesis from glucose. However, Thauer, Jungermann and Decker (1977) commented that the pathway for which acetate and hydrogen is formed from glucose has a very high thermodynamic efficiency (85%), and this is seen to be "incompatible with entropy requirements" and thus butyrate is always a product of acidogenesis, regardless of surge conditions. In their experimental systems, Sam-Soon et al. (1990) did not measure any butyrate production and this was attributed to the fact that butyrate oxidisers become established when the sludge age is longer than 2.3 days. This was seen to be reasonable evidence for excluding the formation of butyrate in the acidogenesis process.

For the acidogenesis process under high hydrogen partial pressure, 1 mole each of propionate, acetate and hydrogen are assumed to form according to (Sam-Soon et al., 1990). This was based on the work of Thauer, Jungermann and Decker (1977). When considering acidogenesis on MS only (as opposed to including AA), a total amount of 66.7% COD is eventually directed to acetate and 33.3% to hydrogen. This is in line with the products formed by glucose acidogenesis and is expected to be consistent across all models since the pathways of glucose utilisation are well-established.

2.1.1.3 Acetogenesis

The methanogenic bacteria are unable to utilise short-chain fatty acids (SCFA) which are higher than C_2 (2 carbon atoms) as their substrate (Thauer, Jungermann & Decker, 1977; Gujer & Zehnder, 1983). This includes propionate, butyrate and valerate. These higher carbon fatty acids are converted to acetate and hydrogen by a group of bacteria called acetogens, by the process called acetogenesis. Different subspecies utilise the different SCFA substrates as the electron donor. Thermodynamically, these reactions are unfavourable at standard temperature and pressure conditions (STP) conditions (Batstone et al., 2002). Accordingly, very low hydrogen partial pressures (and concentrations of other products) are required in order for the forward reaction (production of acetate) to be favoured (Angelidaki et al., 2011). This results in a dependence on the hydrogen-utilising bacteria in order to keep the acetogens operating at their optimal level.

2.1.1.4 Acetoclastic Methanogenesis

The preceding processes form acetate, which is utilised in acetoclastic methanogenesis by the acetoclastic methanogens to form methane (Sötemann et al., 2005b). Approximately 70% of the digester methane is produced from acetate and the remainder from hydrogen. The acetoclastic methanogens are a group of bacteria which are very sensitive to pH and become fully inhibited below pH values of 6 (Batstone et al., 2002) and can only utilise acetate as its' substrate. While typically high ammonia concentrations are inhibitory to bioprocesses, there have been recent discoveries of bacteria which can utilise acetate under high ammonia concentrations (Angelidaki et al., 2011). Inhibition of the acetoclastic methanogens causes accumulation of acetate which further lowers the pH, causing further inhibition of these organisms (Sam-Soon et al., 1990). This is the primary failure mechanism which can cause digester souring – low pH, low H_2CO_3 alkalinity and high VFA concentration (Sötemann et al., 2005b).

2.1.1.5 Hydrogenotrophic Methanogenesis

The hydrogenotrophic methanogens utilise hydrogen (as the electron donor) and carbon dioxide (as the electron acceptor) to form methane. Approximately 30% of methane is produced from hydrogen. Although the hydrogenotrophic methanogens are also pH sensitive like the acetoclastic methanogens, they are only fully inhibited below pH values of 5 (Batstone et al., 2002). Consequently, if digester pH were to decrease, the acetoclastic methanogens would be much sooner inhibited causing an accumulation of acetate, with a further reduction in pH, before the hydrogenotrophic methanogens are inhibited. The importance of the hydrogenotrophic methanogens is to maintain low hydrogen concentrations (or partial pressures) which are critically important in order for the acetogenic bioprocesses to remain favourable (Sam-Soon et al., 1990; Batstone et al., 2002). This interdependent relationship between the hydrogen-utilising bacteria and acetogenic bacteria is defined as syntrophic (Ahring, 2003).

Table 2: Comparison of Reactants Used and Products Formed in AD

Author	Acidogens				Acetogens		Acetoclastic Methanogens		Hydrogenotrophic Methanogens	
	Hydrolysis		Acidogenesis							
	Reactant	Product	Reactant	Products	Reactant	Product	Reactant	Product	Reactant	Product
Mosey (1983)	Not discussed	Not discussed	Glucose	** HBr, HPr, HAc, CO_2 and H_2	HBr, HPr	HAc and H_2	HAc	CO_2 and CH_4	H_2 and CO_2	CH_4 and H_2O
Masse and Droste (2000)	Carbohydrates, lipids, proteins	Soluble sugars, AA, fatty acids	Soluble sugars, AA, fatty acids	HBr, HPr, HAc, CO_2 and H_2	Fatty acids with higher molecular weight than HAc	HAc, CO_2 and H_2	HAc	CO_2 and CH_4	H_2 and CO_2	CO_2 and CH_4
Batstone et al. (2002)	Carbohydrates, lipids, proteins *	MS AA, LCFA	MS AA	Mixed organic acids, CO_2 and H_2	Mixed organic acids (LCFA, HBr, HPr and valerate)	HAc, CO_2 and H_2	HAc (no additional reactants noted)	CO_2 and CH_4	H_2 and CO_2	CO_2 and CH_4
Sötemann et al. (2005b)	Generic Organic - $C_XH_YO_ZN_A$	Glucose	Glucose	HPr, HAc, CO_2 and H_2	HPr	HAc and H_2	HAc	CO_2 and CH_4	H_2 and CO_2	CH_4 and H_2O

* Batstone et al. (2002) mentioned that a step precedes hydrolysis whereby complex organics are first disintegrated to carbohydrates, lipids and proteins

** Mosey (1983) noted that HBr and HPr are only produced as a result of a surge load and HAc is the preferred product.

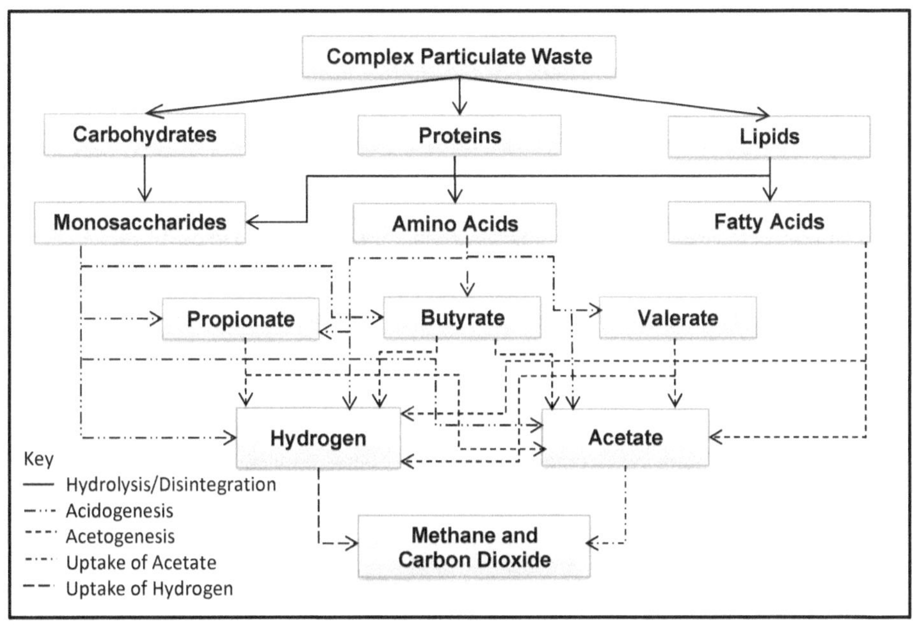

Figure 1: Reactants and Products of Organism Groups in ADM1

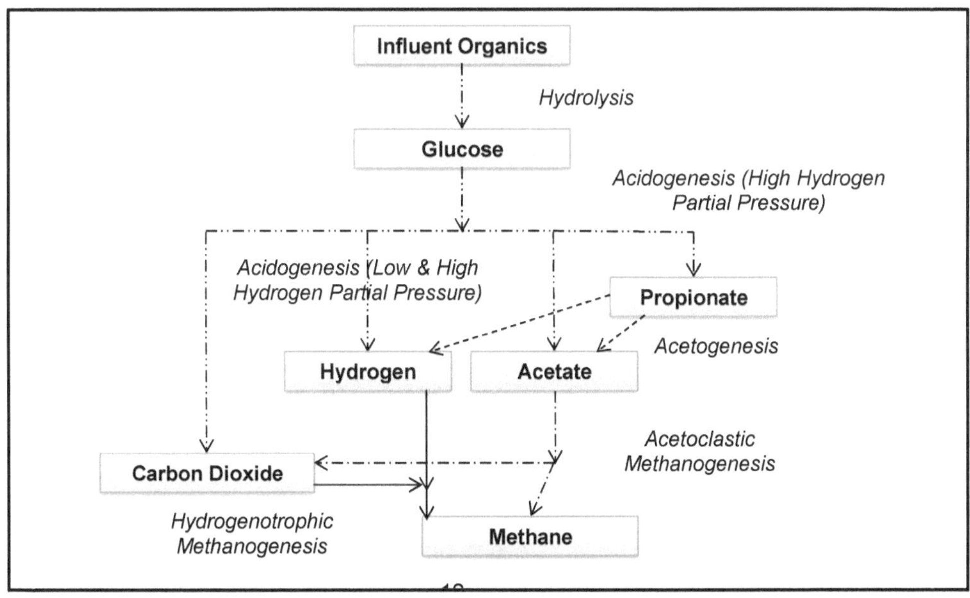

Figure 2: Reactants and Products of each Organism Group According to Söteman et al. (2005)

Adopted from Mosey (1983)

Figure 2 above summarises the products and reactants of each process as described by Sötemann et al. (2005b), Sam-Soon et al. (1990) and Mosey (1983) for the bioprocesses following hydrolysis, while Figure 1 summarises the products and reactants for ADM1 (Batstone et al., 2002).

2.1.2 Kinetic Rates of Hydrolysis of Sewage Sludge

Hydrolysis of the BPO in an anaerobic environment was identified to be the rate-limiting step in AD of sewage sludge by Sötemann et al. (2005b), Sötemann et al. (2005a), Ristow et al. (2005) and Lyberatos and Skiadas (1999) amongst others because the organics comprise mostly particulate organics.

Ristow et al. (2005), Sötemann et al. (2005a) and Sötemann et al. (2005b) all mentioned that four kinetic equations could be considered when evaluating the rate of hydrolysis namely: first-order kinetics, first-order specific kinetics, saturation kinetics and Monod Kinetics. These four rate expressions have all been used to model various biological processes in the past in activated sludge (AS) and AD (Sötemann et al., 2005a), and are discussed in more detail below.

First-order kinetics is a rate formulation which describes the rate of an equation which is dependent on only one reactant. Furthermore, the rate is proportional only to the selected reactant. When applied to the rate of hydrolysis of BPO in AD, the rate of the process is dependent only on the residual BPO COD concentration, S_{bp} (Sötemann et al., 2005a) and ignores the biomass concentration that may be mediating the hydrolysis. This equation has previously been used to describe the hydrolysis of sewage sludge in AD by Henze and Harremoës (1983), Bryers (1985) and Vavilin et al. (2001). The rate expression is illustrated in Equation (1) below.

$$r_h = K_h S_{bp} \tag{1}$$

where

r_h is the volumetric rate of hydrolysis [gCOD/(L.day)]

K_h is the first order hydrolysis rate constant (gCOD/L)

S_{bp} is the biodegradable particulate organics concentration (gCOD/L)

First-order specific kinetics includes the effect of the acidogen biomass concentration (Z_{AD}) on the rate of hydrolysis and thus is included in the rate expression, together with the BPO COD concentration (Sötemann et al., 2005b). It assumes that the acidogens mediate the process of hydrolysis (Sötemann et al., 2005a). This formulation has been used to model the conversion of readily biodegradable organics to VFAs in anaerobic reactors of enhanced biological phosphorus removal (EBPR) systems by Wentzel et al. (1988) and in Activated Sludge Model No. 2 (ASM2) (Henze et al., 2000). The rate expression is illustrated in Equation (2) below.

$$r_h = K_H S_{bp} Z_{AD} \qquad (2)$$

where

r_h is the volumetric rate of hydrolysis [gCOD/(L.day)]

K_H is the first order specific hydrolysis rate constant [L/(gCOD.day)]

S_{bp} is the biodegradable particulate organics concentration (gCOD/L)

Z_{AD} is the acidogen active biomass concentration (gCOD/L)

Monod Kinetics is widely used in modelling bioprocesses in biological wastewater treatment. It has been used in activated sludge models to describe the rate of the utilisation of ammonia by autotrophic nitrifiers and readily biodegradable organics (Sötemann et al., 2005a). The rate expression is illustrated in Equation 3.

$$r_h = \frac{K_M S_{bp}}{K_S + S_{bp}} Z_{AD} \qquad (3)$$

where

r_h is the volumetric rate of hydrolysis [gCOD/(L.day)]

K_M is the maximum specific hydrolysis rate constant [gCOD/(gCOD.day)]

K_S is the half-saturation constant for hydrolysis (gCOD/L)

S_{bp} is the biodegradable particulate organics concentration (gCOD/L)

Z_{AD} is the acidogen active biomass concentration (gCOD/L)

Lastly, saturation kinetics (also known as Contois (1959) kinetics) have been used to model the hydrolysis of complex organics in activated sludge models by Dold, Ekama and Marais (1980) who used the planar-surface mediated reaction kinetics from Levenspiel (1972). Since hydrolysis in AS systems operates similarly to hydrolysis in AD systems, saturation kinetics were investigated for describing the rate of hydrolysis in AD as well Sötemann et al. (2005b). The rate expression is given by Equation 4 below.

$$r_h = \frac{K_M \left(\frac{S_{bp}}{Z_{AD}}\right)}{K_S + \left(\frac{S_{bp}}{Z_{AD}}\right)} Z_{AD} \tag{4}$$

where:

r_h is the volumetric rate of hydrolysis [gCOD/(L.day)]

K_M is the maximum specific hydrolysis rate constant [gCOD/(gCOD.day)]

K_S is the half-saturation constant for hydrolysis (gCOD/L)

S_{bp} is the biodegradable particulate organics concentration (gCOD/L)

Z_{AD} is the acidogen active biomass concentration (gCOD/L)

The saturation kinetics equation is similar in form to the Monod kinetics equation in that at low S_{BP}/Z_{AD}, the rate is first order with respect to S_{BP}/Z_{AD}, and at high S_{BP}/Z_{AD}, it is zero-order with respect to S_{BP}/Z_{AD}. The important difference between Monod and saturation kinetics is that with Monod kinetics, the rate is dependent on the bulk liquid concentration of S_{BP} and independent of the S_{BP}/Z_{AD}, food to microorganism ratio (F/M) whereas in saturation kinetics, the rate is dependent on the S_{BP}/Z_{AD}, F/M ratio and independent on bulk liquid S_{BP} concentration. This difference is crucially important for digester start-up where the F/M ratio is high but the bulk liquid S_{BP} and Z_{AD} concentrations are very low. Saturation kinetics was introduced into activated sludge models to make the kinetic rate independent of the bulk liquid S_{BP} concentration so that the BOD test and aerated lagoons could be modelled with the same AS kinetics (Dold, Ekama & Marais, 1980).

Ristow et al. (2005), Sötemann et al. (2005a) and Sötemann et al. (2005b) completed extensive work to determine which hydrolysis rate equation was superior for use in AD models. Their results are summarised in Table 3 below, which shows that authors clearly differ in their preferred rate formulation. Depending on the dataset, the type of sludge as well as the digester SRT, the different rate formulations predict different results. The choice of the hydrolysis kinetic rate equation depends on the model application.

Sötemann et al. (2005a) concluded that for primary sludge (PS), once calibrated, each kinetic equation was found to be equally useful for checking the COD removal and gas production for different retention times in steady-state applications because each kinetic equation was calibrated to the same PS dataset. While each kinetic equation yields slightly different results for the unbiodegradable COD fraction (f'_{PSup}) of the PS, the difference in the f'_{PSup} was very small, ranging between 0.33 and 0.36. They also found that pure PS digested significantly faster than a mixture of PS and humus sludge (trickling filter biofilm) resulting in better COD removals and higher gas production at the same sludge age for the pure PS.

For the hydrolysis of cellulose material which has a large component of particulates, Vavilin, Rytov and Lokshina (1996) found that the Monod equation predicted the process the worst. This was interpreted to mean that the hydrolysis process cannot be considered an enzymatic reaction. First-order kinetics, while it gives reasonably good predictions in the literature, was found to be inadequate for short sludge ages where large amounts of biomass may be washed out of the anaerobic digester resulting in less biomass being present therein to mediate the hydrolysis process, resulting in a slowing down of the rate (Vavilin, Rytov & Lokshina, 1996). Incorporating saturation kinetics allow for better predictions over a broader range of operating sludge ages (Sötemann et al., 2005b).

Table 3: Comparison of Hydrolysis Kinetic Rate Equations

Author	Preferred Hydrolysis Rate	Data Used	Additional Comments
Sötemann et al. (2005a)	Monod	Izzett (1992), O'Rourke (1967)	Once calibrated, found that either equation is acceptable with steady-state models at long SRT (>7days)
Sötemann et al. (2005b)	Saturation	Izzett (1992), O'Rourke (1967)	First-order and saturation kinetics were seen to be more superior to Monod and first-order specific kinetics. Saturation kinetics includes the acidogen organism group within the equation. Thus it is intuitively more superior than first order and also has been applied successfully in activated sludge models such as Activated Sludge Model No. 1 (ASM1).
Ristow et al. (2005)	First Order Kinetics	Ristow et al. (2005), Izzett (1992), O'Rourke (1967)	For steady-state conditions at long sludge ages, once calibrated all kinetic equations yield virtually identical results. Large percentage of errors were obtained with Monod and first-order specific kinetics. Saturation kinetics resulted in low percentage errors, but first order was selected instead due to the accuracy and simplicity associated with this kinetic equation.

2.1.3 Digester Inhibition and Failure

Toxicity and inhibition have previously been described by Batstone et al. (2002) as an "adverse effect on bacterial metabolism" and an "impairment of bacterial function" respectively. Chen, Cheng and Creamer (2008) state that a substance is regarded as inhibitory if it causes an "adverse shift in the microbial population". The acetoclastic methanogens are one of the most important groups of bacteria in AD which can significantly impact digester stability if inhibited.

Several toxins have been identified in AD, namely sulphide, light metal ions and heavy metals. Chen, Cheng and Creamer (2008) discuss details of these toxins. Presence of one or more inhibitory substances can cause digester upset, which is indicated by a reduction in gas yield or decrease in pH, and lack of response to this instability can cause irrecoverable digester failure. Toxins need not only be already present in the substrate but could also be generated during the degradation process such as VFA's, ammonia as NH_3 and sulphide. Toxic substances which are already in the substrate include heavy metals and antibiotics (Boe, 2006).

Process failure is most commonly associated with a rise in VFA, which is usually associated with a drop in pH resulting in further inhibition. If steps are not taken to remedy the situation (by raising the pH), complete, and generally, irreversible failure takes place (Li et al., 2017). This phenomenon is also sometimes termed digester "souring". The methanogens and the acetogens are the key organism groups required to keep the digester stable, and suitable inhibition terms are thus necessary in AD models to ensure that they are represented accurately.

While high concentrations of ammonia are inhibitory to the acetoclastic methanogens, a different organism group termed the acetate oxidisers develop under high ammonia concentration conditions. These organisms oxidise acetate to hydrogen, which is then utilised by the hydrogenotrophic methanogens to form methane (Wett et al., 2014). These anaerobic organism groups are more dominant when operating thermophilic reactors, or in AD treating concentrated sewage sludge (~10%TSS) subjected to high pressure (~6 atm) and temperature (~150°C) pre-treatment.

2.1.3.1 pH Inhibition of the Methanogens

The bacteria which produce acetate (acidogens and acetogens) and those that consume acetate (methanogens) differ widely in their preferred temperature, pH, kinetic growth rates and sensitivity to environmental conditions. The balance between these bacteria is critical to maintaining a stable digester (Chen, Cheng & Creamer, 2008). Acetoclastic methanogens thrive at near-neutral pH conditions and are fully inhibited below a pH of 6 (Batstone et al., 2002). Sam-Soon et al. (1987) found that preventing the pH from falling below 6.6 in a UASB reactor, which is in a failed state in the bottom, allowed recovery to take place further up the bed. This phenomenon was cited by Moosbrugger et al. (1993b) as well. However, others have found that the acetoclastic methanogens have an optimal pH range between 6.6 and 7.2 (Siegrist et al., 2002) or 7.3 and are strongly inhibited below 6.2 (Pereira et al., 2013).

The inhibition function used in ADM1 (Batstone et al., 2002) has the acetoclastic methanogens fully inhibited below 6, and fully functional above a pH of 7. Although the acetoclastic methanogens do not thrive well under pH values greater than 8 (Pereira et al., 2013), it is uncommon for the digester pH to reach such values and thus inhibition at low pH values only is seen to be sufficient. The pH inhibition function of acetoclastic methanogens (AM) is given by Equation (5) and shown graphically in Figure 3. For the acetoclastic methanogens, the upper limit (pH_{UL}) is 7, and the lower limit (pH_{LL}) is 6. As discussed above, although the pH inhibition of the hydrogenotrophic methanogens is unlikely to cause the failure, their inhibition is included with Equation (5) with pH_{UL} of 6 and a pH_{LL} of 5.

$$I = \begin{cases} e^{-3\left(\dfrac{pH - pH_{UL}}{pH_{UL} - pH_{LL}}\right)^2}, & pH < pH_{UL} \\ 1, & pH \geq pH_{UL} \end{cases} \tag{5}$$

In contrast, the inhibition function for acetoclastic methanogens proposed by Sötemann et al. (2005b) - shown below in Equation (6) and also in Figure 3 - has a broader range of operation at lower pH values and at the optimal pH of 7, are 92% active. Furthermore, the K_I value of 1.15×10^{-6} mol/L translates to a pH of 5.93. Based on the definition of the inhibition constant, the activity of the methanogens is at 50% at a pH of 5.93, which is in sharp contrast to that proposed by other sources. Increasing the inhibition constant, K_I in Equation (6), allows the organisms to be more active (>92%) at the optimal pH value, but because of the form of the function, it also allows them to be more active at the lower pH values.

$$I = \frac{K_I}{K_I + [H^+]}$$

(6)

where

K_I is the inhibition constant = 1.15 x 10^{-6} mol/L (the value at which the activity of the methanogens is half the normal rate)

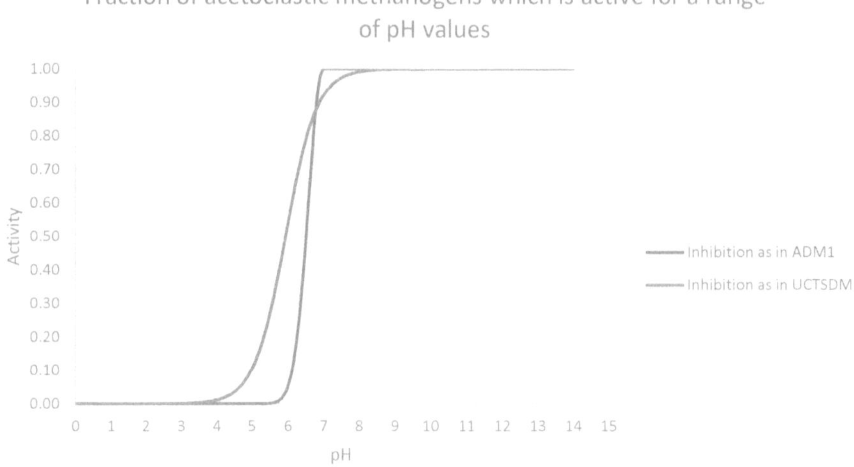

Figure 3: Graphical Representation of Inhibition on Acetoclastic Methanogens due to pH

2.1.3.2 Hydrogen Inhibition of the Acetogens

According to Thauer, Jungermann and Decker (1977), the reactions by which the SCFA (larger than C_2) form acetate and hydrogen for the methanogens to use, are unfavourable at STP conditions. This is shown in the reaction below by the positive free energy of the reaction at standard conditions ($G^{0'}$). STP conditions refer to standard temperature and pressure which is 25°C (expressed in Kelvin as 298), gas pressures of 1 atm, concentrations of 1 M and also a pH of 7 which is indicated by the prime mark in $\Delta G^{0'}$ (Thauer, Jungermann & Decker, 1977).

$$CH_3CH_2COO^- + 3H_2O \rightarrow HCO_3^- + CH_3COO^- + H^+ + 3H_2 \quad \Delta G^{0'} = +\frac{76.1kJ}{reaction}$$

If this reaction were to proceed spontaneously, $\Delta G'$ would need to be negative. Therefore, the reaction products concentrations need to be significantly lower than that at STP conditions (<<<1M) and substantially lower than the reactants. An example of this (as given by Batstone et al. (2002)) is that if the partial pressure of hydrogen (H_2) is less than 10^{-5} atm at a temperature of 298K, it will result in $\Delta G'$ being negative which allows a spontaneous reaction provided that the remaining product concentrations are not too large.

$\Delta G'$ is calculated with Equation (7) with the gas constant R = 8.314J/mol/K and the temperature in Kelvin.

$$\Delta G' = \Delta G^{0'} + RTln\left(\frac{[C]^c[D]^d}{[A]^a[B]^b}\right) \tag{7}$$

Where A and B are reactants and C and D are products in a reversible reaction with coefficients a, b, c and d respectively:

$$aA + bB \leftrightarrow cC + dD$$

For simplicity, aqueous hydrogen concentrations are normally used in modelling, not partial pressures (Batstone et al., 2002). A non-competitive function was proposed in ADM1 and inhibitory hydrogen concentrations of 1 x 10^{-6} kgCOD/m^3, corresponding to 7 x 10^{-5} bar, was found for the propionate-utilising bacteria. This was based on the assumption of gas-liquid equilibrium. Björnsson, Murto and Mattiasson (2000) report the inhibitory dissolved hydrogen concentration to be greater than 0.04µM which translates to a COD concentration of 0.64 x 10^{-6} kgCOD/m^3, 0.64 x 10^{-3} mgCOD/L or 0.08 x 10^{-3} mgH$_2$/L.

Sam-Soon et al. (1990) reported the acetogens to be sensitive to hydrogen partial pressures above $10^{-4.1}$ atm but the temperature for which this assumption was made is not given. In subsequent work where Sam-Soon et al. (1991b) developed a mathematical model for a glucose-fed UASB reactor, the value for the half-saturation constant in the switching function was selected to be 10mgCOD/l, which seems very high. This value is intrinsically linked to the growth rate of acetogens and calibration on this process may require the value to be refined.

2.1.4 Warning Indicators

While AD has been used for a long time, they are often operated far below the maximum capacity to avoid overloading, failure and very costly recovery. One of the significant limitations to overcome this problem is the absence of robust control mechanisms that adequately represent the dynamics of the AD process to avoid failure. Failure is that point at which methane production is limited, and pH has fallen (causing inhibition of the acetoclastic methanogens) such that the influent feed has to be suspended. Available controlling techniques are generally very complicated, labour intensive, time-consuming or do not give sufficient early warning to prevent anaerobic digester failure. Typically, VFAs, alkalinity, pH, and gas production and composition are used to analyse anaerobic digester stability (Björnsson, Murto & Mattiasson, 2000), and multiple parameters give complementary information (Hickey et al., 1991). Even a generally low-loaded anaerobic digester can experience an imbalance between VFA production and utilisation resulting in a low pH and inhibition of AM when loaded with readily biodegradable organics. The issue of anaerobic digester stability, therefore, has more to do with keeping the consortium of organisms synchronised at low or high-loaded anaerobic digesters. Björnsson, Murto and Mattiasson (2000) discuss that the lack of reliable online methods affects the choice of the monitoring method. This section discusses the various warning indicators from literature and its suitability to identify impending digester failure.

2.1.4.1 Gas-Phase Indicators

Monitoring of gas production is usually used for evaluating AD performance. However, the gas production was found to be the least sensitive indicator of impending digester failure likely due to the limitations in the mass transfer of liquid to gas (Li et al., 2014). Others have also reported that a large headspace results in a slow response of gas composition which may not allow gas monitoring to be used successfully for process imbalance but may provide information on process performance (Boe et al., 2010). Consequently, because gas flow and composition are not found to be suitable for AD instability monitoring, dissolved hydrogen concentrations can be measured instead (Björnsson et al., 2001).

2.1.4.2 pH

pH is a simple parameter to measure in the digester utilising a pH probe. Since it is a direct measurement, the information is quickly available without any complicated titrations and time delays. Very often, it is the only online measurement taken. It is a very widely used indicator of anaerobic digester stability, but it may not be a reliable indicator when the buffer capacity or H_2CO_3 alkalinity of the aqueous phase is high. Björnsson, Murto and Mattiasson (2000) found good system response with pH limit of 6.8 as the indicator for process stress since the buffer capacity in the lab-scale anaerobic digester was low but in general, pH as an indicator by itself is not sufficient because the buffer capacity can vary. It is recommended therefore that pH, as well as either Partial Alkalinity (PA) or VFA, be monitored on-line. In contrast, Franke-Whittle et al. (2014) found that with the lab-scale reactors operated, highly buffered reactors did not experience a low pH with accumulation in VFA, but at the same time, these reactors also did not fail. This confirmed that different anaerobic digesters could withstand different loads of VFA prior to failure, as stated by Ahring, Sandberg and Angelidaki (1995).

2.1.4.3 VFA

Since the primary mechanism of AD failure is an accumulation of VFA, which decreases H_2CO_3 alkalinity, whether it is as a result of a toxin, overloading or inhibition of the AM organism group, it seemed beneficial to develop an easier method to determine these two concentrations for simpler anaerobic digester control. Previous methods of estimating total acetate species require expensive equipment and considerable skill. These methods include distillation, gas chromatography, high-pressure liquid chromatography, Fourier transform infrared spectroscopy (Hey et al., 2013) or straight distillation which may not give accurate results (Moosbrugger et al., 1993c).

Moosbrugger et al. (1993d) developed the 5-point titration which allows the H_2CO_3 alkalinity, and VFA (or SCFA) to be determined with known concentrations of phosphate (Ekama & Brouckaert, 2019), ammonia and sulphide. This method is robust, user-friendly, requires simple equipment and simultaneously determines both H_2CO_3 alkalinity as well as VFA concentration. Usually, the H_2CO_3 alkalinity is high enough to ignore the small alkalinity contributions of the ammonia, phosphate and sulphide weak acid/base systems, provided these FSA, OP and sulphide concentrations are low enough. This has led to it receiving significant attention (Poinapen, Ekama & Wentzel, 2009a; Hey et al., 2013). Though the calculations by hand are not practical, software such as Titra5 allows the estimates to be completed with ease, which reduces this method to a simple titration of no more effort than the PA and Total Alkalinity (TA) titration but yields much better information (Ekama et al., 2019).

Subsequent to the development of the 5-point titration, an 8-point titration has also been developed, however Hey et al. (2013) found that due to the high buffer capacity in the low pH range, which the extra 3 pH points required to be titrated to, less accurate VFA concentrations were determined for the primary sludge hydrolysis case.

The 5-point titration thus provides a tool for the VFA as well as H_2CO_3 alkalinity to be determined online (Hey et al., 2013). However, as discussed above, accumulation of VFA to the point at which it is inhibitory differs for each system which is why it cannot be used as a stand-alone indicator. A significant advantage of having determined the VFA is to use it to determine the standard AD stability indicator, the Ripley ratio (Ripley, Boyle & Converse, 1986).

2.1.4.4 Alkalinity

The TA can be defined as the proton accepting capacity between an initial pH and another equivalence lower pH point. The proton donating capacity is negative alkalinity which is called acidity (Moosbrugger et al., 1993a). TA is expressed with respect to reference species of each weak acid/base significantly present and is the sum of all the weak acid/base system species that can be associated with an H^+ to become reference species (Loewenthal, Ekama & Marais, 1989; Loewenthal et al., 1991) - see Equation (8a). The TA includes the VFA as well as carbonate species, which are considered to be the main buffering agents in AD (Björnsson et al., 2001) because the phosphate and sulphide are generally not significantly present and the pK value of ammonia is too high (9.1) to add significantly to the TA. Since both carbonate and VFA are included in the TA, this parameter has been found to be insensitive to process instability since VFA accumulates and increases, bicarbonate alkalinity (BA) decreases, but TA may increase or remain constant depending on the magnitude of VFA increase (Björnsson et al., 2001).

$$Total\ Alk = Alk\ H_2CO_3 + Alk\ H_3PO_4 + Alk\ HAc + Alk\ NH_4{}^+ + Alk\ H_2S$$
$$+ Alk\ H_2O$$
$$= [HCO_3{}^-] + 2[CO_3{}^{2-}] + [H_2PO_4{}^-] + 2[HPO_4{}^{2-}] + 3[PO_4{}^{3-}] +$$
$$[CH_3COO^-] + [NH_3] + [HS^-] + 2[S^{2-}] + [OH^-] - [H^+]$$

(8a)

(8b)

PA is assumed to give a good representation of BA and is found by titrating to a pH endpoint of 5.75. While Intermediate Alkalinity (IA) is assumed to give a good approximation of the VFA concentration and is found by titrating to a pH of 4.3, this is not the case – much more accurate estimates of the VFA and H_2CO_3 alkalinity concentrations are obtained with the 5-point titration for the same effort even if the OP, sulphide and FSA concentrations are assumed zero (Ekama et al., 2019).

Combined indicators relying on alkalinity such as VFA/TA, BA/TA, IA/PA (or equivalently VFA/BA) and the Ripley ratio (IA/PA) have been reported to give a better early-warning time compared with single indicators such as pH, gas production and concentrations of propionate and acetate (Li et al., 2017). The indicator Pr/Ac was found to give a warning that is too late to avoid impending digester failure because propionate accumulation does not always occur before acetate accumulation. Since ammonia adds to the available, total alkalinity when at high concentration, it can buffer against an increase in VFA concentrations, thus, AD at high ammonia concentration will result in reduced sensitivity of indicators which are dependent on the BA. It was for this reason that Li et al. (2014) found that an AD reactor at high concentration of ammonia was most sensitive to the VFA/TA.

2.2 ANAEROBIC DIGESTION MODELLING

Generally, models are used to represent dynamic conditions, where changes occur as a function of time (Henze et al., 2008). Ekama (2009) stated that dynamic models contain differential equations (DEs) of many bioprocesses, and numerical integration is required in order to solve these equations. In contrast, steady-state conditions prevail when the system operates under conditions of constant influent flow and load and the operating conditions are held constant (Billing & Dold, 1988). Steady-state models are simpler to use and implement, and they also demand much less input information than dynamic models. Moreover, Ekama (2009) mentioned that in steady-state models, most of the bioprocesses, except the slowest one, are assumed to reach completion and thus can be resolved stoichiometrically.

Henze et al. (2008) lists the several advantages of mathematical modelling. Some of these include being able to gain insight into plant performance, evaluate possible upgrading and new plant designs and provide operator training. Importantly, Henze et al. (2008) highlights that dynamic models are used to apply sensitivity analysis and optimise the steady-state design.

To date, several AD models have been developed. These models have provided varying degrees of insight into the interaction between biological reactions of the different organism groups (Gujer & Zehnder, 1983; Mosey, 1983) and the weak acid-base chemistry of the aqueous phase in which they function. The interaction with the aqueous phase via pH was recognised early on due to the sensitivity of the methanogens to pH and so was included in early AD models but not in an explicit and generalised way due to the complexity of the interaction (Graef & Andrews, 1974; McCarty, 1974). More recent AD models, such as International Water Association (IWA) Anaerobic Digestion Model No 1 (ADM1) developed by Batstone et al. (2002) and UCTSDM developed by Sötemann et al. (2005b), have accounted for this interaction in a more generalized way using separated bioprocess stoichiometry and thermodynamic equilibria based on mass and charge balance approaches (Musvoto et al., 1997; Musvoto, Ekama, et al., 2000; Musvoto, Wentzel & Ekama, 2000; Musvoto, Wentzel, et al., 2000) .

The UCTSDM is based on the granulation UASB reactor AD model of Sam-Soon et al. (1990) who based their AD model on that of Mosey (1983). Mosey (1983) introduced the first AD model that integrates the activities of four AD organism groups; 1) acid-forming bacteria, 2) acetogenic bacteria, 3) acetoclastic methanogen, and 4) hydrogen-utilising methanogen groups and included the effect of hydrogen (H_2) concentration and redox potential as key parameters that regulate the activity of some of these groups. The Mosey (1983) model also considered the impact of overloading on the activities of the microbial groups and on the performance of the system.

However, the Mosey (1983) AD model is based on glucose as feed organics, which was useful in the UASB reactor research treating various concentrated soluble industrial organic wastes (Sam-Soon et al., 1991a; Moosbrugger et al., 1993b,e; Wentzel et al., 1994). It is the reason why in the UCTSDM the hydrolysis of complex organics bioprocess produces glucose as an intermediate product (Sötemann et al., 2005b) - it allowed using the Mosey approach for modelling the four AD organism groups following hydrolysis of complex organics instead of hydrolysis producing carbohydrates, lipids and proteins as is done in IWA ADM1.

Based on the two-phase (aqueous-gas) UCTSDM, Brouckaert, Ikumi and Ekama (2010); Ikumi (2011); Ikumi, Brouckaert and Ekama, 2011; Ikumi et al. (2015)   developed PWM_SA_AD by adding:

- Phosphorus within the organics and biomass compositions, polyphosphate in phosphorus accumulating organisms (PAO) and ortho-P speciation in the aqueous phase;

- Multiple organic groups (VFA, dissolved, non-settleable and settleable biodegradable and unbiodegradable organics and biomass) each with their own $C_xH_yO_zN_aP_b$ composition;

- Three-phase (aqueous-gas-solid) mixed weak acid/base chemistry and multiple mineral precipitation; and

- Separated fast aqueous phase speciation equilibrium processes with algebraic equations (AE) and slow bioprocess, gas exchange and precipitation processes with kinetic DEs.

Further details of this model are given by Ikumi et al. (2015). Table 4 compares the main advantages and disadvantages of the three AD models discussed above.

While the different AD models have their strengths and weaknesses, most have not been fully calibrated for all the AD organism groups – calibration depended on the type of organics modelled. Earlier AD models by Graef and Andrews (1974) and Hill and Barth (1977), considered methanogenesis to be the rate-limiting bioprocess. In contrast, Sötemann et al.

(2005a), Ikumi et al. (2011), (2015) and Ristow et al. (2005) considered the hydrolysis of complex organics to be rate-limiting, which is reasonable for an anaerobic digester which is fed PS and waste activated sludge (WAS). While this is satisfactory for AD in plant-wide and water resource recovery facility (WRRF) situations where AD dynamics are limited, it disqualifies UCTSDM for dynamic conditions where there may be more than one rate-limiting bioprocess depending on the dynamic loading conditions. Therefore, in its present state, UCTSDM and PWM_SA_AD cannot be used to simulate the dynamics of digester failure because the kinetics of the bioprocesses following hydrolysis has not been calibrated.

Table 4: Summary of AD Models Advantages and Disadvantages

AD Model	Advantages	Disadvantages
IWA ADM1	• The slow biological processes are separated from the fast ionic equilibrium equations by external AE – the model is not "stiff".	• Carbohydrates, proteins and lipids are not routinely measured – substrate is difficult to characterise. • Charge balance - all aqueous ions need to be accounted for.
UCTSDM	• The substrate is characterised using typical wastewater measurements. • It includes an ion-pairing effect on pH. • It includes elemental mass balances of CHON and COD.	• Calculation of pH is done internally. This results in model stiffness, slow runtimes and possible numerical instability. • Calibrated only for the hydrolysis process of PS and WAS.
PWM_SA_AD	• The substrate is characterised using typical wastewater measurements. • Separation of fast and slow processes prevents solver instability and reduces runtime • Is a three-phase model and thus gas exchange and precipitation reactions can be modelled. • Includes pK value correction based on ionic strength as well as ion pairing. • Includes elemental mass balances of CHONP and COD. • Uses an aligned measurement and modelling framework that allows measurement of the ionic strength via conductivity, routine wastewater analytical methods and mixed weak acid/base chemistry principles.	• Calibrated only for the hydrolysis process of PS and WAS.

Because of its roots in the Mosey (1983) AD model, PWM_SA_AD provides the best basis for the further development of an AD model.

2.3   UPFLOW ANAEROBIC SLUDGE BED (UASB) REACTORS

High-rate anaerobic treatment systems in upflow anaerobic sludge bed (UASB) reactors gained popularity because of their sludge retention feature which allowed low HRT (Lettinga et al., 1984; Sam-Soon et al., 1987; Moosbrugger et al., 1993d; Wentzel et al., 1994). With some readily biodegradable organics like glucose, a granular sludge bed formed and the large granule size (3-5mm) ensured fast settling. The UASB reactor, therefore, can withstand high inflow velocities and short HRT (Sam-Soon et al., 1990).

Also, in AD, when a high hydrogen partial pressure is present, acidogenesis produces propionate as an additional product (Sam-Soon et al., 1987) as shown in Figure 2. In UASB reactor systems, this is generally observed with dissolved readily biodegradable organics because UASB reactor systems are subject to loading rates which are two to three times higher than loading rates applied to completely-mixed anaerobic digesters (Sam-Soon et al., 1987). In addition, due to the solids accumulation at the bottom of the UASB reactor, reactants are used at very high rates resulting in a rapid accumulation of products. This results in a build-up of acetate and propionate in the bottom part of the bed since the acetogens can only utilise propionate to form acetate once the hydrogen partial pressure decreases below $10^{-4.1}$ atm (Sam-Soon et al., 1990).

Usually, in high rate UASB reactors, the bottom part of the bed is in a state of failure due to the very fast production rates of acetate and propionate from the readily biodegradable organics feed. However, the system as a whole does not fail because the readily biodegradable organics of the feed are depleted before permanent failure takes place, allowing the upper part of the bed to recover by utilising the acetate and propionate, which raises pH (Sam-Soon et al., 1990). In operating granular UASB reactors it is important to keep the pH above 6.6 in the bed otherwise the acetoclastic methanogens become too inhibited for the upper part of the bed to recover (Moosbrugger et al., 1993e) and the 5- point titration (Moosbrugger et al., 1993d) was developed to facilitate UASB reactor operation and alkalinity dosing to control and maintain the pH above 6.6.

Once the substrate has been utilised in the bottom part of the bed, the system has a chance to recover in the upper levels of the bed. Recovery commences when a low hydrogen partial pressure is restored, and the propionate can commence being utilised by the acetogens (or propionate degrading bacteria). In addition, acetate also accumulates in the lower part of the bed since the large accumulation of acetate and propionate causes low pH levels. The acetoclastic methanogens are therefore temporarily slowed in the low pH range but no so much such that the acetate is not utilised which is why the lowest pH in the bed should not fall below 6.6. Once the system recovers, resulting in an increase in pH, the acetate can again be utilised at a fast rate (Sam-Soon et al., 1987).

Due to these dynamics, the rates at which the acetate and propionate are consumed are important because they are not utilised as fast as they are formed. This temporary instability and failure condition at the bottom of the UASB reactor forms the basis for investigating anaerobic digester failure and developing a model to predict digester failure, with a suitable dataset.

Identification of a suitable dataset for investigating digester instability involved reviewing various UASB reactor experimental systems which were operated under varying conditions. Summaries of the available literature of UASB reactor experimental systems can be found in Dutta, Davies and Ikumi (2018). The experimental systems were grouped based on either the substrate fed, the temperature at which they were operated, mixing conditions or pH variation within the UASB reactors. Following the review of the experimental systems, the lack of data for bed measurements of intermediate products was highlighted. Many of the datasets focussed on gas flow measurements at the outflow and have influent characterisation, without any measurements along the reactor height. Two datasets, however, had measurements along the bed height for the typical AD intermediate products, namely the dataset of Sam-Soon (1989) and Kalyuzhnyi et al. (1996). The latter dataset's applicability was however excluded due to the accumulation of butyrate instead of propionate – which is not included in the model at this stage.

2.4 THE SANI PROCESS

2.4.1 Background

Hong Kong has had a dual water supply system since 1957 with seawater being supplied for flushing toilets and potable water is used for other domestic purposes. The implementation of a dual water supply system has resulted in 270 million m^3 of fresh water being saved annually (or equivalently, 740Ml/day) (Water Supplies Department, 2012).

Flushing toilets with seawater results in the generation of saline sewage of about ⅓rd seawater salinity. The concentrations are on average 5000mg/L chloride, 400mg/COD/L and 500 mg/L sulphate, with a typical COD/Sulphate (as S) ratio of 2.4 (Lu et al., 2009; Wang et al., 2009, 2011). For sulphate reduction to occur, 2g of COD is required per gram of sulphate (as S). Since the typical saline sewage characteristics result in a COD/S ratio of 2.4, the sewage can be treated anaerobically utilising sulphate-reducing bacteria (SRB).

Apart from the shortage of fresh water in Hong Kong, the remaining capacity within the landfills is an additional problem. Anaerobic treatment of sewage with sulphate-reducing bacteria produce significantly less sludge compared with the aerobic treatment of sewage with AS (Wang et al., 2009). Additionally, Wang et al. (2009) indicates that by using sulphate as an electron acceptor, low sludge yield can be expected compared with using nitrate and oxygen as electron acceptors. Conversion of FSA to nitrogen gas is required to prevent eutrophication in downstream water bodies. Typically, this is accomplished through heterotrophic denitrification and autotrophic nitrification. Sludge production in heterotrophic denitrification and organics removal is much higher (0.4gVSS/COD) compared with autotrophic denitrification (0.1gVSS/COD) (Wang et al., 2009). Based on the above rationale of treating saline sewage biologically, while still removing nitrogen with low sludge yields, the SANI process was developed (Lu et al., 2009; Wang et al., 2009, 2011; Lu et al., 2012b).

2.4.2 SANI Process Details

The SANI process allows for the removal of organics and with sulphate-reducing bacteria, autotrophic denitrification (using dissolved sulphide as the electron donor and nitrate as the electron acceptor) and nitrification, to allow for the conversion of FSA to nitrate.

Figure 4 shows a schematic layout of the SANI system. BSR and organics removal occur in the SRUSB by SRB. The dissolved sulphide generated in the anaerobic reactor is then used as an electron donor in the anoxic reactor for autotrophic denitrification, where nitrate is converted to nitrogen gas. Lastly, an aerobic zone nitrifies the ammonia to nitrate, and the nitrate is recycled to the anoxic reactor. The aerobic zone also oxidises any remaining sulphide to sulphate.

A key issue in this system is that any excess sulphide produced not required for removal of nitrate is oxidised back to sulphate with oxygen which is a waste of energy – instead greater removal of organics should be achieved in the primary separation unit (PSU) before the SRUSB resulting in a higher methane production in the methanogenic AD of the primary sludge (PS). Operating the SANI system at the optimum organics removal in the PSU for minimum energy consumption with aeration and maximum energy generation in the SRUSB will yield a net energy-producing WRRF.

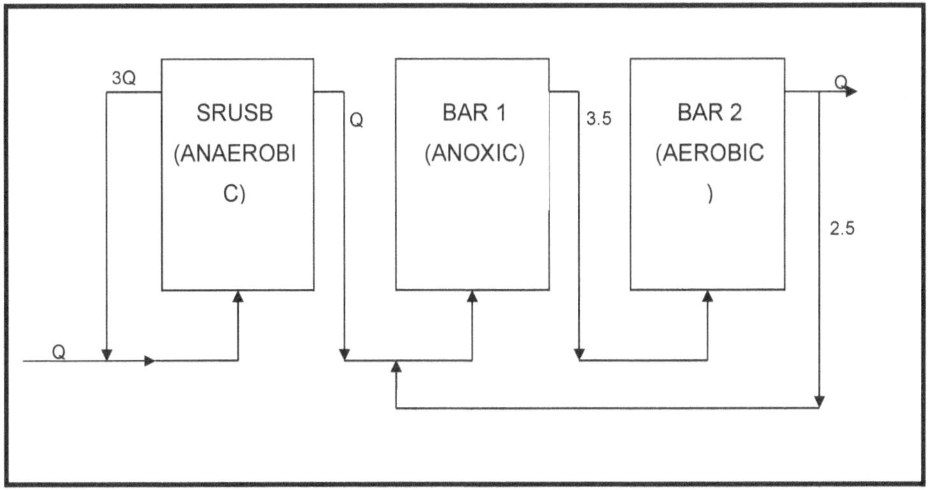

Figure 4: Simplified Diagram of the SANI Process   Adapted from Lu *et al.* (2012)

As mentioned above, the SANI system includes an anaerobic SRUSB where organics removal and sulphate reduction occur producing sulphide. Due to the absence of methanogens, the BSR bioprocesses in the SRUSB are mediated by parallel SRB which is similar to methanogenic anaerobic bioprocesses. A summary of the organism groups mediating the BSR processes is shown in Figure 5. Additional details can be found in Hao et al. (2014). Adding BSR to the PWM_SA_AD model should not present major difficulties – the three methanogenic organism groups that follow acidogenesis are replaced with similar sulphidogenic organism groups.

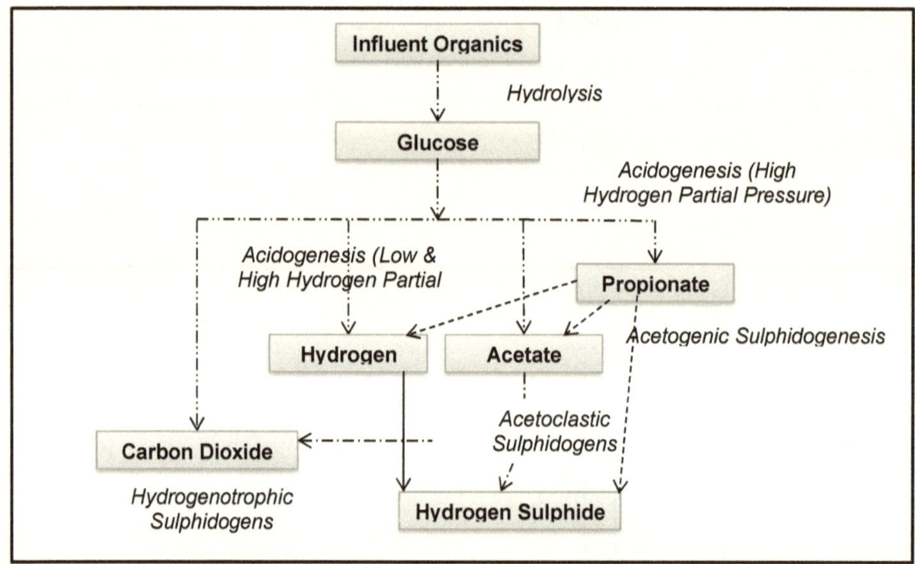

Figure 5: Products and Reactants in Biological Sulphate Reduction

Because the SANI system employs an SRUSB (UASB) reactor for the purposes of sulphate reduction, a UASB model would be required to model the SRUSB reactors with separated HRT and SRT (Poinapen, Wentzel & Ekama, 2009). Sludge retention in the UASB requires good solid-liquid separation, which depends on the sludge settleability and bed expansion. Poinapen, Ekama and Wentzel (2009b) assessed the settleability and bed expansion of the sludge by means of a settleometer which measured the settling velocity of the solids fraction sizes in the sludge bed. By means of these measurements, they were able to assess whether bed expansion dynamics or sludge settling velocity was the dictating parameter for capacity determination. Also, "bed mixing" to prevent sludge flotation was investigated by Wang et al. (2017) and they determined that gas recirculation creating a shear gradient of 4.2s^{-1} resulted in the lowest sludge flotation potential well below the limit to cause bed flotation. Therefore, the smallest reactor short-circuiting flow has the lowest dead zone volume and the lowest energy consumption compared with hydraulic or mechanical mixing.

To date, only completely-mixed and membrane AD models have been developed (Batstone et al., 2002; Sötemann et al., 2005b; van Zyl, 2008). Although both Sam-Soon et al., (1991) and Poinapen and Ekama (2010a) developed models for the UASBs which they operated, they were modelled as single completely-mixed reactors of volume equal to the bed volume and a point settler (or membrane) for (perfect) solid-liquid separation. These models were therefore not capable of modelling the internal bed dynamics such as the transitionary failed state of the UASB and its recovery.

It would appear from the observed behaviour of the SANI system SRUSB, in which granulation has been observed to take place (Hao et al., 2015) like in methanogenic UASBs (Sam-Soon et al., 1990), that due to the sludge accumulation at the bottom of the bed, hydrolysis of BPO cannot any longer be assumed to be the rate-limiting bioprocess. Incomplete utilisation of the influent BSO at the short bed HRT may result in high effluent dissolved organics which disrupt the subsequent autotrophic MBBR. Therefore, successful modelling of the SRUSB requires at least a model that can simulate a UASB which has been further calibrated. The calibration protocol developed in, this book could be extended for the SRUSB which will enhance the modelling outcome of the SANI process.

2.5 CONCLUSION

The literature discussed above provided a background into AD modelling and highlighted the organism groups present as well as the substrates utilised, and products formed of each one. One of the significant differences in AD modelling between various models and studies is the assumed rate-limiting step. PWM_SA_AD was identified to be the most suitable AD model for this purpose.

While the hydrolysis step may be rate-limiting under many conditions especially if municipal primary sludge is being fed to the digester, under conditions leading to failure, this may no longer be true. This motivates the need for suitable calibration of the model under such conditions, and UASBs provide the ideal place to begin such calibration as they have a build-up of intermediate products in the bottom of the bed.

In conditions approaching failure, the accumulation of intermediate products may give insight into which measurement to use as a warning indicator. The selection of the warning indicator also depends on its ease of frequent measurement. From the literature reviewed, it is apparent that combined indicators, such as the Ripley ratio, may give better warnings than single indicators, such as pH.

Lastly, the SRUSB within the SANI system requires a calibrated UASB model for improved modelling outcomes. This is due to the high number of solids in the bottom of the bed which results in rapid utilisation of particulate material, but the high inflow rate can cause soluble organics to rise up the bed and exit the SRUSB, entering into the anoxic MBBR and thereby hindering autotrophic denitrification.

3 MODEL DESCRIPTION

As concluded in the Literature Review above, PWM_SA is shown to be the most appropriate model for the purposes of developing a UASB AD model. This section provides an overview of the model, beginning with the model terminology used for ease of understanding and presents in detail the structure and contents of the model.

3.1 TERMINOLOGY

Some of the common terminology used within the model is presented here in order to allow a concise understanding of the model details. The notation developed by Corominas et al. (2010) was implemented in the model in order for ease of understanding between model developers.

3.1.1 Components

Components are specific combinations of elements (and charge) – they specialise the stoichiometry for a particular system and are subject to the material content of the system. The choice of the components for PWM_SA_AD, listed in Table 5, was made to represent all unit operations universally across the plant-wide PWM_SA model. Each component is parameterised based on its COD and Molar concentrations (Molality). Importantly, the components are not specific values but are a group of defined names. Their composition is determined by so-called parameters, which define the component, and is either fixed for known components (like PO_4^{3-} and NH_4^+) or editable for variable components (such as FBSO or BPO). Hydrogen gas (H_2) is assumed to be dissolved, and its composition is fixed because it is known that one mole of it comprises two atoms of hydrogen.

The editable component compositions are for the organic material such as influent BPO (denoted as X_{Binf}) where its composition can vary significantly at different WRRF. The composition of these components is then specified by assigning its parameter values. Therefore Table 6 indicates the composition of these editable components such as $CH_YO_ZN_AP_BS_C$. Based on the notation of Corominas et al. (2010), Y, which is the subscript of the H (hydrogen) element is i_H_X_{Binf}_mol_perC which now becomes easy to understand because the name is self-explanatory. The composition of organics and biomass was extended to include S (sulphur) if required, although in this book the value was set to 0.

The following components were added to PWM_SA:

- HS⁻ (sulphide)

- X_ACS (acetogenic sulphidogens)

- X_AS (acetoclastic sulphidogens)

- X_HS (hydrogenotrophic sulphidogens)

- X_ADO (autotrophic denitrifying organisms). These organisms, like heterotrophic denitrifiers, produce nitrogen gas. However, in the model, the nitrogen which is produced in the bioprocesses is modelled as dissolved nitrogen (S_N_2) and not as gaseous nitrogen.

- Nitrogen gas (G_N_2) was also added as a component so that the headspace of an anaerobic digester can be purged with N_2 gas prior to start-up; it does not feature in any bioprocesses but simply exits the system as the headspace gets occupied by the gases formed by the bioprocesses.

Table 5: Model components used in PWM_SA (Brouckaert, Ikumi & Ekama, 2010)

	Component Name	Empirical formula	Notation
Solvent	Water	H_2O	H_2O
Total Dissolved Ionic Concentrations	Hydrogen ion	H^+	S_H
	Sodium	Na^+	S_Na
	Potassium	K^+	S_K
	Calcium	Ca^{2+}	S_Ca
	Magnesium	Mg^{2+}	S_Mg
	Ammonium	NH_4^+	S_NH_x
	Chloride	Cl^-	S_Cl
	Acetate	CH_3COO^-	S_VFA
	Propionate	$CH_3CH_2COO^-$	S_Pr
	Carbonate	CO_3^{2-}	S_CO_3
	Sulphate	SO_4^{2-}	S_SO_4
	Phosphate	PO_4^{3-}	S_PO_4
	Bisulphide	HS^-	S_HS
	Nitrate	NO_3^-	S_NO_x
Soluble Component	Dissolved hydrogen	H_2	S_H_2
	Unbiodegradable Soluble Organics	$CH_{Yu}O_{Zu}N_{Au}P_{Bu}S_{Cu}$	S_U

	Component Name	Empirical formula	Notation
	Fermentable Biodegradable Soluble Organics	$CH_{Yf}O_{Zf}N_{Af}P_{Bf}S_{Cf}$	S_F
	Glucose	$C_6H_{12}O_6$	S_Glu
	Dissolved oxygen	O_2	S_O
	Dissolved nitrogen	N_2	S_N2
Particulates	Unbiodegradable particulate organics	$CH_{Yup}O_{Zup}N_{Aup}P_{Bup}S_{Cup}$	X_U_inf
	Biodegradable particulate organics	$CH_{Ybp}O_{zbp}N_{Abp}P_{Bbp}S_{Cbp}$	X_B_Org
	Primary sludge biodegradable particulate organics	$CH_{Ybps}O_{Zbps}N_{Abps}P_{Bbps}S_{Cbps}$	X_B_Inf
	Polyphosphate	$K_{kp}Mg_{mp}Ca_{cp}PO_3$	X_PAO_PP
	Poly-hydroxy-alkanoate	$C_4H_6O_2$	X_PAOtor
	Struvite	$MgNH_4PO_4.6H_2O$	X_Str_NH4
	Calcium Phosphate	$Ca_3(PO_4)_2$	X_ACP
	K-struvite	$MgKPO_4.6H_2O$	X_Str_K
	Calcite	$CaCO3$	X_Cal
	Magnesite	$MgCO3$	X_Mag
	Newberyite	$MgHPO4$	X_Newb
	Influent inorganic settleable solids		X_ISS
Microorganism Biomass	Ordinary heterotrophic organisms (OHOs)	$CH_{Yo}O_{Zo}N_{Ao}P_{Bo}S_{Co}$	X_OHO
	Phosphate accumulating organisms	$CH_{Yo}O_{Zo}N_{Ao}P_{Bo}S_{Co}$	X_PAO
	Autotrophic denitrifying organisms	$CH_{Yo}O_{Zo}N_{Ao}P_{Bo}S_{Co}$	X_ADO
	Autotrophic nitrifying organisms	$CH_{Yo}O_{Zo}N_{Ao}P_{Bo}S_{Co}$	X_ANO
	Acidogens	$CH_{Yo}O_{Zo}N_{Ao}P_{Bo}S_{Co}$	X_AD
	Acetogens	$CH_{Yo}O_{Zo}N_{Ao}P_{Bo}S_{Co}$	X_AC
	Acetogenic sulphidogens	$CH_{Yo}O_{Zo}N_{Ao}P_{Bo}S_{Co}$	X_ACS
	Acetoclastic methanogens	$CH_{Yo}O_{Zo}N_{Ao}P_{Bo}S_{Co}$	X_AM
	Acetoclastic sulphidogens	$CH_{Yo}O_{Zo}N_{Ao}P_{Bo}S_{Co}$	X_AS
	Hydrogenotrophic methanogens	$CH_{Yo}O_{Zo}N_{Ao}P_{Bo}S_{Co}$	X_HM
	Hydrogenotrophic sulphidogens	$CH_{Yo}O_{Zo}N_{Ao}P_{Bo}S_{Co}$	X_HS
	Endogenous residue	$CH_{Ye}O_{Ze}N_{Ae}P_{Be}S_{Ce}$	X_U_Org
Gases	Carbon dioxide	CO_2	G_CO2
	Methane	CH_4	G_CH4
	Nitrogen gas	N_2	G_N2

3.1.2 Parameters

The parameters are the model constants which can be obtained from (i) measurement in an experimental system, (ii) calibration with a model or (iii) from literature. Examples of parameters include the kinetic rate constants, temperature, elemental stoichiometric composition values and yield coefficients. The parameters may be changed between simulation runs if required through the user-interface, but they are held constant during a simulation. Parameters can also be coded in as equations which are calculated at the beginning of the simulation. An example of this would be the electron-donating capacity (EDC) of the organisms (γ_o – gam_o) which is determined from the biomass composition values. Because its value does not change during a simulation, it is considered a parameter rather than a variable. This is the main difference between parameters and variables. Table 6 provides a list of the parameters as well as their default values prior to calibration.

Table 6: List of Parameters of PWM_SA_AD

	Parameter	Description	Value	Unit
System Parameters	V_liq	The volume of liquid in the reactor	System-specific	m^3
	V_gas	The volume of gas in the reactor		m^3
	Temperature	System temperature		°C
	Tref	Reference Temperature	35	°C
	TempCoeff	Rate Temperature Coefficient	0.0667	-
	MM_C	The molar mass of carbon	12.011	g/mol
	MM_H	The molar mass of hydrogen	1.0079	g/mol
	MM_O	The molar mass of oxygen	15.999	g/mol
	MM_N	The molar mass of nitrogen	14.007	g/mol
	MM_P	The molar mass of phosphorus	30.974	g/mol
	MM_S	The molar mass of sulphur	32.065	g/mol
	MW_[Comp_Index]	Array components providing the molar mass of each	Vector parameters	g/mol
	COD_per_mol_[Comp_Index]	Array components providing the COD of each		gCOD/mol
Stoichiometric parameters	*gam_o	Electron donating capacity (EDC) of organism mass	Calculated parameters	e⁻/mol
	*gam_f	EDC of fermentable organics		e⁻/mol
	*gam_e	EDC of endogenous residue		e⁻/mol
	*gam_bp	EDC of the biodegradable particulate portion of biomass decay		e⁻/mol
	*gam_bps	EDC of the biodegradable particulate influent organics		e⁻/mol
	*gam_us	EDC of the unbiodegradable soluble organics		e⁻/mol
	*gam_up	EDC of the unbiodegradable particulate organics		e⁻/mol
	Stoi_H2O_decay	Coefficient of water in the decay process		g/mol
	Stoi_H_decay	Coefficient of H^+ in the decay process		g/mol
	Stoi_CO3_decay	Coefficient of CO_3^{2-} in the decay process		g/mol
	Stoi_NH4_decay	Coefficient of NH_4^+ in the decay process		g/mol
	Stoi_PO4_decay	Coefficient of PO_4^{3-} in the decay process		g/mol
	Stoi_SO4_decay	Coefficient of SO_4^{2-} in the decay process		g/mol
	Stoi_ER_decay	Coefficient of X_U_Org in the decay process		g/mol

Parameter	Description	Value	Unit
f_XU_Bio_lysis	The unbiodegradable fraction of biomass that accumulates on lysis with death regeneration model	0.08	-
Y_AC	Acetogenesis biomass yield	0.0397	molOrg/molsub
Y_AD	Lo H2 Acidogenesis biomass yield	0.0895	molOrg/molsub
Y_AH	Hi H2 Acidogenesis biomass yield	0.0895	molOrg/molsub
Y_AM	Acetoclastic Methanogenesis biomass yield	0.03925	molOrg/molsub
Y_HM	Hydrogenotrophic Methanogenesis biomass yield	0.04	molOrg/molsub
i_H_Org_mol_perC	H/C : organisms	1.4	dUnit/dUnit
i_HF_mol_perC	H/C : fermentable soluble	2.01	dUnit/dUnit
i_HU_mol_perC	H/C: unbiodegradable soluble	1.646	dUnit/dUnit
i_H_XBInf_mol_perC	H/C: PS biodegradable particulate	2.19	dUnit/dUnit
i_H_XBOrg_mol_perC	H/C: biodegradable particulate	1.463	dUnit/dUnit
i_H_XUInf_mol_perC	H/C: unbiodegradable particulate	1.482	dUnit/dUnit
i_H_XUOrg_mol_perC	H/C: endogenous residue	1.482	dUnit/dUnit
i_K_PP_mol_perP	Molar fraction of K/P in polyphosphate	0.31506	dUnit/dUnit
i_Mg_PP_mol_perP	Molar fraction of Mg/P in polyphosphate	0.31669	dUnit/dUnit
i_N_Org_mol_perC	N/C : organisms	0.2	dUnit/dUnit
i_N_SF_mol_perC	N/C: fermentable soluble	0.119	dUnit/dUnit
i_N_SU_mol_perC	N/C: unbiodegradable soluble	0.062	dUnit/dUnit
i_N_XBInf_mol_perC	N/C: PS biodegradable particulate	0.0643	dUnit/dUnit
i_N_XBOrg_mol_perC	N/C: biodegradable particulate	0.229	dUnit/dUnit
i_N_XUInf_mol_perC	N/C: unbiodegradable particulate	0.113	dUnit/dUnit
i_N_XUOrg_mol_perC	N/C: endogenous residue	0.113	dUnit/dUnit
i_O_Org_mol_perC	O/C : organisms	0.4	dUnit/dUnit
i_O_SF_mol_perC	O/C : fermentable soluble	0.592	dUnit/dUnit
i_O_SU_mol_perC	O/C: unbiodegradable soluble	0.593	dUnit/dUnit
i_O_XBInf_mol_perC	O/C: PS biodegradable particulate	0.653	dUnit/dUnit
i_O_XBOrg_mol_perC	O/C: biodegradable particulate	0.355	dUnit/dUnit
i_O_XUInf_mol_perC	O/C: unbiodegradable particulate	0.472	dUnit/dUnit
i_O_XUOrg_mol_perC	O/C: endogenous residue	0.472	dUnit/dUnit
i_P_Org_mol_perC	P/C : organisms	0	dUnit/dUnit
i_P_SF_mol_perC	P/C: fermentable soluble	0.012	dUnit/dUnit
i_P_SU_mol_perC	P/C: unbiodegradable soluble	0.02	dUnit/dUnit
i_P_XBInf_mol_perC	P/C: PS biodegradable particulate	0.0097	dUnit/dUnit
i_P_XBOrg_mol_perC	P/C: biodegradable particulate	0.031	dUnit/dUnit
i_P_XUInf_mol_perC	P/C: unbiodegradable particulate	0.022	dUnit/dUnit
i_P_XUOrg_mol_perC	P/C: endogenous residue	0.022	dUnit/dUnit
i_S_Org_mol_perC	S/C : organisms	0	dUnit/dUnit
i_S_SF_mol_perC	S/C: fermentable soluble	0	dUnit/dUnit

Parameter	Description	Value	Unit
i_S_SU_mol_perC	S/C: unbiodegradable soluble	0	dUnit/dUnit
i_S_XBInf_mol_perC	S/C: PS biodegradable particulate	0	dUnit/dUnit
i_S_XBOrg_mol_perC	S/C: biodegradable particulate	0	dUnit/dUnit
i_S_XUInf_mol_perC	S/C: unbiodegradable particulate	0	dUnit/dUnit
i_S_XUOrg_mol_perC	S/C: endogenous residue	0	dUnit/dUnit
KS_AC	Half saturation coefficient for acetogens	0.089	g/m^3
KS_AD	Half saturation coefficient for acidogens	0.78	g/m^3
KS_AM	Half saturation coefficient for acetoclastic methanogens	0.013	g/m^3
KS_HM	Half saturation coefficient for hydrogenotrophic methanogens	0.156	g/m^3
KS_BInf_AD_hyd	Half saturation coefficient for X_B_Inf (Primary sludge)	10.124	gCOD/gCOD
KS_BOrg_AD_hyd	Half saturation coefficient for X_B_Org (organics from biomass death)	10.37	gCOD/gCOD
K_CO2	Rate constant for CO2 exchange	0.1	1/d
K_I_H2	Inhibition coefficient for H2 in acidogenesis	1.25	g/m^3
K_I_H_AM	H+ inhibition for acetoclastic methanogens	1.15E-06	mol/kg
K_I_H_HM	H+ inhibition for hydrogenotrophic methanogens	0.00053	mol/kg
b_AC	Decay rate constant for acetogens	0.015	1/d
b_AD	Decay rate constant for acidogens	0.041	1/d
b_AM	Decay rate constant for acetoclastic methanogens	0.037	1/d
b_HM	Decay rate constant for hydrogenotrophic methanogens	0.01	1/d
b_OHO_AD	Decay rate constant for OHOs in AD	20	1/d
b_PAO_AD	Decay rate constant for PAOs in AD	20	1/d
kH_F_AD_hyd	Hydrolysis rate constant for FBSO	10	1/d
kH_PHA_AD_hyd	Hydrolysis rate constant for poly-hydroxy-alkanoate (PHA)	5	1/d
kH_PP_AD_hyd	Hydrolysis rate constant for polyphosphate (PP)	1	1/d
kM_BInf_AD_hyd	Hydrolysis rate constant for X_B_Inf	2.004	1/d
kM_BOrg_AD_hyd	Hydrolysis rate constant for X_B_org	1.95	1/d
kM_fPP_PAO_PHAstor	Maximum rate for PP release with anaerobic PHA storage	0.03	1/d
mu_AC (equivalent to μ_{AC})	Max specific growth rate for acetogens	1.15	1/d
mu_AD (equivalent to μ_{AD})	Max specific growth rate for acidogens	0.8	1/d
mu_AM (equivalent to μ_{AM})	Max specific growth rate for acetoclastic methanogens	4.39	1/d
mu_HM (equivalent to μ_{HM})	Max specific growth rate for hydrogenotrophic methanogens	1.2	1/d

The items marked with * in Table 6 are calculated with Equation (9) below, which is the EDC (γ_o or gam_o – used in WEST ®). The EDC of the different organism groups is calculated with the same Equation (9) but the terms in the equation change to those relevant to the specific component. So, the EDC of the organisms (gam_o) is shown in Equation (10) where the terms inserted into Equation (10) are all related to organisms (Org). When calculating γ (EDC) of fermentable soluble organics (S_F), gam_o becomes gam_f, and "Org" in the parameters becomes "SF" e.g., i_H_Org_mol_perC becomes i_H_SF_mol_perC.

$$\gamma = 4k + l - 2m - 3n + 5p + 6s \tag{9}$$

Where the biomass is represented by $C_kH_lO_mN_nP_pS_s$ and k is assigned a value of 1.

$$gam\_o = 4 + i\_H\_Org\_mol\_perC - 2 \times i\_O\_Org\_mol\_perC - 3 \times \tag{10}$$
$$i\_N\_Org\_mol\_perC + 5 \times i\_P\_Org\_mol\_perC + 6 \times i\_S\_Org\_mol\_perC$$

3.1.3 Variables

Variables are components or species which change at each time step during a simulation. Within WEST®, they are divided into Input, Output, Derived and Algebraic variable groups. Derived variables include the initial digester conditions and component masses which need to be specified within a reasonable range of the final simulation values prior to simulation. These initial masses allow the initial component and species concentrations to be calculated, which are then fed into the kinetic expressions to determine the kinetic rates of the bioprocesses at the initial and subsequent time steps.

3.1.4 Species and Speciation Routine

In this extended version of PWM_SA of which PWM_SA_AD is a subset, there are 46 ionic species and complexes linked to a total of 13 ionic components which align with the dissolved components from Table 5. Species are the subgroups of these 13 dissolved components, for example, S_CO3 represents the total carbonate concentration in the system as a component, and this comprises H_2CO_3, HCO_3^-, CO_3^{2-} species as well as other carbonate complex species as a result of ion-pairing such as $MgCO_3$ and $CaHCO_3^+$ which are determined by the speciation routine. The species that were selected to be components in the bioprocess stoichiometry was based on those species adopted in well-known physicochemical models (PHREEQC (Parkhurst & Appelo, 2013)and MINTEQA2 (Allison, Brown & Novo-Gradac, 1991)) for ease of cross-referencing and model checking.

Table 7: Ionic Components, Species and Complex Ion Pairs

Dissolved Ionic Component	Empirical Formula of Component	Ionic Species and Complexes
S_H	H^+	H^+, OH^-, HCO_3^-, H_2CO_3, NH_3, HPO_4^{2-}, $H_2PO_4^-$, H_3PO_4, HAc, HPr, $CaOH^+$, $MgOH^+$, $NaHCO_3$, $NaHPO_4^-$, $CaHCO_3^+$, $CaHPO_4$, $MgHCO_3^+$, $MgHPO_4$, $MgH_2PO_4^+$
S_Na	Na^+	Na^+, $NaHPO_4^-$, $NaCO_3^-$, $NaHCO_3$, NaAc, $NaSO_4^-$
S_K	K^+	K^+
S_Ca	Ca^{2+}	Ca^{2+}, $CaCO_3$, $CaHCO_3^+$, $CaPO_4^-$, $CaHPO_4$, $CaSO_4$, $CaOH^+$, $CaAc^+$, $CaPr^+$
S_Mg	Mg^{2+}	Mg^{2+}, $MgCO_3$, $MgHCO_3^+$, $MgPO_4^-$, $MgHPO_4$, $MgH_2PO_4^+$, $MgSO_4$, $MgOH^+$, $MgAc^+$, $MgPr^+$
S_NH	NH_4^+	NH_4^+, NH_3, $NH_4SO_4^-$
S_Cl	Cl^-	Cl^-
S_VFA	CH_3COO^-	Ac^-, HAc, $CaAc^+$, NaAc, $MgAc^+$
S_Pr	$CH_3CH_2COO^-$	Pr^-, HPr, $CaPr^+$, $MgPr^+$
S_CO3	CO_3^{2-}	CO_3^{2-}, HCO_3^-, H_2CO_3, $CaCO_3$, $MgCO_3$, $CaHCO_3^+$, $MgHCO_3^+$, $NaCO_3^-$, $NaHCO_3$
SO4	SO_4^{2-}	SO_4^{2-}, $CaSO_4$, $MgSO_4$, $NH_4SO_4^-$, $NaSO_4^-$
S_PO4	PO_4^{3-}	PO_4^{3-}, HPO_4^{2-}, $H_2PO_4^-$, H_3PO_4, $MgPO_4^-$, $CaPO_4^-$, $MgHPO_4$, $CaHPO_4$, $NaHPO_4^-$, $MgH_2PO_4^+$
S_HS	HS^-	HS^-, H_2S

At a time-step of a simulation, the DEs comprise the stoichiometry and kinetics of the slow physical (gas exchange and precipitation), and bioprocesses are solved resulting in the component values at the next time step. The new component values are then speciated into the species concentrations with an algebraic equation speciation routine. The speciation equations are not time-based because the endpoint of these extremely fast aqueous phase equilibrium processes is defined by the thermodynamics of the dissociation processes.

This approach keeps separate the very fast aqueous equilibrium processes and slow physical and biological processes for fast runtimes and mathematically stable solver calculations. This approach may initially seem confusing because some species are also components (e.g., CO_3^{2-} and PO_4^{3-}), but the context makes clear what the variable is – in the DEs, they are components, and in the speciation routine, they are species.

3.2 MODEL DESCRIPTION

3.2.1 Overview

PWM_SA_AD is a subset of the 3-phase (aqueous-gas-solid) full element (C, H, O, N, P), COD and charge mass balance plant-wide WRRF (PWM_SA) model developed by Brouckaert, Ikumi and Ekama (2010) and Ikumi et al. (2015). As a subset of PWM_SA, PWM_SA_AD includes all the features of PWM_SA. This PWM_SA model combines biological N and P removal AS, AD of primary sludge and AD or anoxic-aerobic digestion of WAS with interlinking non-reactive thickening physical unit operations. It defines influent wastewater organics concentrations in the same seven organics groups as in municipal wastewater (VFA, BSO and BPO, USO and UPO) each in the generic $C_xH_yO_zN_aP_b$ form (as shown in Table 5 above) and uses routinely measured parameters (e.g. COD, VSS, TOC, TKN, TP) to quantify the x,y,z,a,b values and is full element, COD and charge mass balanced. Within this study, the compositions of the organic have been extended to include sulphur if required, so now the generic organics form is $C_xH_yO_zN_aP_bS_c$.

An external algebraic equation equilibrium speciation sub-routine allows amongst other aspects the pH, gas partial pressure, ionic concentrations and ionic strength to be calculated. Since it separates the slow biological processes from the faster aqueous processes, it improves model stiffness. It uses the non-ideal determination of dissociation constants and includes ion pairing, which allows modelling of integrated physical, chemical and biological processes systems at high ionic strength (<0.7 mol/kg)

Finally, because the model is a three-phase model, it includes the effect of mineral precipitation on the pH and ionic strength of the system due to the release of phosphate, ammonia and inorganic carbon (IC) species in the AD system. These features of the PWM_SA model are therefore also included in its PWM_SA_AD sub-model. One of the limitations of the PWM_SA_AD model is that it was calibrated only for the rate-limiting hydrolysis bioprocess in the conventional AD of sewage sludge (Ikumi, Harding & Ekama, 2014).

The plant-wide model (PWM_SA) is based on the "super-model" approach of Jones and Takacs (2004) and Seco et al. (2004). The approach allows direct mapping of the processes in ASM2 (Gujer et al., 1995), and PWM_SA_AD on an elemental component base. This has the advantage of

(i)   placing the physicochemical states globally and linking the biological components between the AS (which is based on ASM2) and AD subsets of the model,

(ii)   including parameterised stoichiometry (the x, y, z, a, b, c values of the influent organics groups and biomass species) for the bioprocesses,

(iii)   sharing the same ionic speciation in all reactive unit operations and as a result

(iv)   the output components of the ASM2 subset become the input components directly for the AD subset without the need for transformation equations.

In general, the simulations of the different physical, chemical and biochemical processes in PWM_SA are based on determining the materials present at a particular location and time (mass balancing) and determining the physical state that it will take on at that point (speciation) (Brouckaert, Ikumi & Ekama, 2010). In order to perform these calculations, the PWM_SA adopted a general physicochemical modelling framework (Brouckaert, Brouckaert & Ekama, 2019a). The hierarchy of the entities considered in the model and their interrelationships are illustrated in Figure 6. The PWM_SA model is given in the WRRF simulation software platform WEST®. Further detail of this model is given by Ikumi, Harding and Ekama (2014) and Ikumi et al. (2015).

3.2.2 PWM_SA Physico-Chemical Model Framework

The principal objective of developing this approach is to accommodate the different rates of reactions in the WRRF/AD system that can vary enormously. Accordingly, PWM_SA and its subset PWM_SA_AD incorporates a framework that enables the model to separate the fast and slow biological processes.

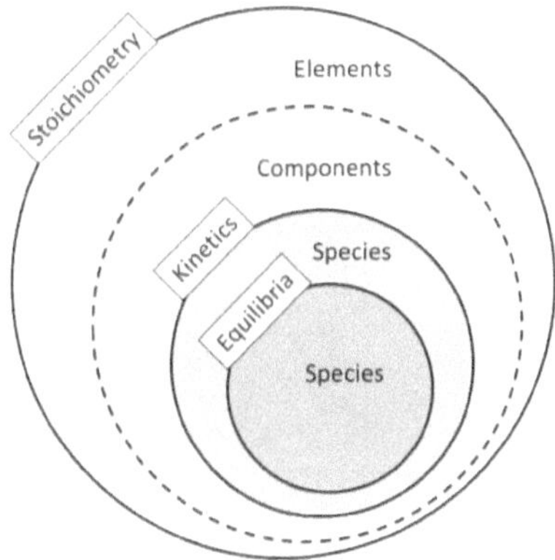

Figure 6: Physico-chemical modelling framework

Figure 6 shows that mass balance of material content (stoichiometry – outer circle) applies throughout and the masses of the elements (C, H, O, N, P, S and charge) represent this material content.

Species are those material entities which are required to describe the actual physical state of the material content. Kinetic considerations add physical detail and require species to describe the time-dependent behaviour of the system. The kinetic constants in these rates define the endpoint of the biological and physical processes that do not reach equilibrium – these processes require calibration and the kinetic constants hold these processes away from equilibrium, a state which differs for different system constraints imposed on these processes. The final endpoints of the kinetic processes that do reach equilibrium are defined by the thermodynamics of the processes. Kinetic relationships still apply to these processes, but the driving force becomes zero very quickly. Because the endpoint of the very fast equilibrium processes is defined by their thermodynamics, which is well-defined, these processes usually do not need calibration.

3.2.3 Biological Reactions

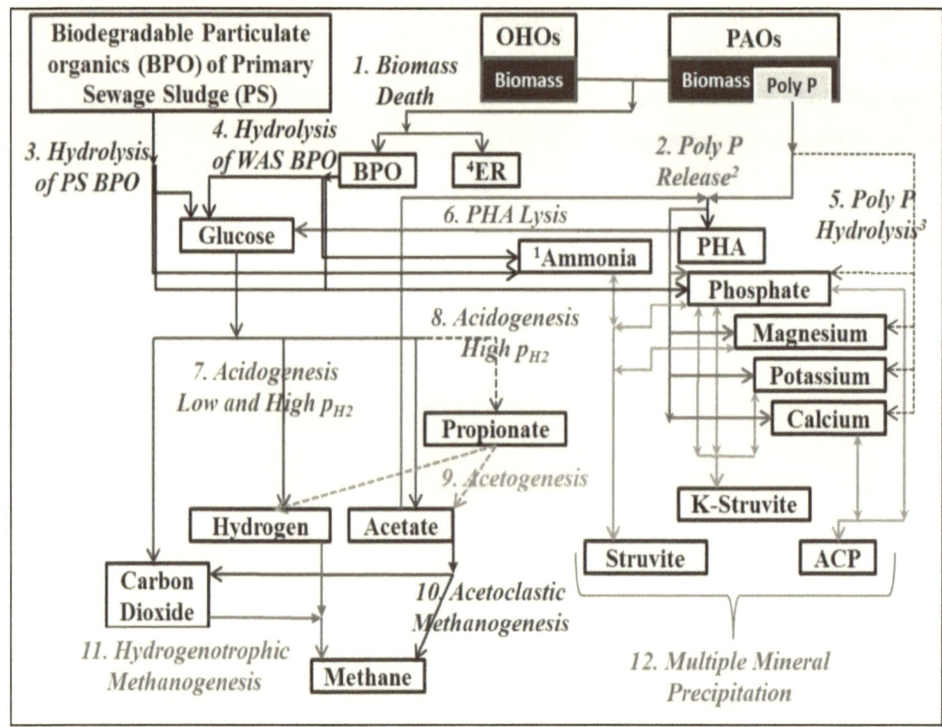

Figure 7:Overview of the PWM_SA Three-phase Process Scheme   Adapted from Ikumi (2011)

Figure 7 illustrates the schematic of the general AD biological and physical precipitation processes and the different organism groups included in the PWM_SA_AD. Although in Table 5 it is shown that the set of components representing micro-organism groups all have the same stoichiometric composition, they are grouped in Figure 7 based on the biological reactions they mediate - having the same stoichiometric composition does not render these biomass groups independent from the mass balance (stoichiometry) but they are kinetically independent and so are treated as model components, rather than as species. The stoichiometric dependence does make it difficult to determine parameters associated with these biomass components from experimental data but has an advantage of making the model relatively easy to expand and include processes that are not currently included in the model.

The process schematic for PWM_SA_AD extended the 2-phase (aqueous-gas) AD model of Sötemann et al. (2005b) to the 3-phase (aqueous- gas-solid) AD model (Brouckaert, Ikumi & Ekama, 2010; Ikumi, 2011). However, the bioprocess stoichiometry of PWM_SA (AS and AD subsets) was derived differently to Sötemann et al. (2005b) by considering the electron donor reaction and the anabolic and catabolic electron destination reactions separately and adding them to obtain the CHONPS and charge mass balanced stoichiometric equations (Ekama et al., 2019). The biomass yield coefficient values of Sötemann et al. (2005b) therefore needed to be changed from mol biomass/mol substrate to gCOD biomass/gCOD substrate. As shown in Figure 7, PWM_SA_AD assumes a single hydrolysis bioprocess for the feed sewage sludge (primary and/or waste activated sludge) into the idealised intermediate product glucose. This approach was chosen because the biochemical pathways of AD of glucose are well known, and so the AD model of Mosey (1983) could be directly used. Further, from Mosey (1983) and Sam-Soon et al. (1990), the model includes the effects of high hydrogen partial pressure on the acidogenic and acetogenic bioprocesses to produce not only acetate and hydrogen (under low hydrogen partial pressure) but additionally propionate under high hydrogen partial pressure with the propionate utilisation taking place under low hydrogen partial pressure only.

At this stage of AD model development, hydrogen is modelled as a dissolved concentration rather than a gas due to complications with modelling the gas evolution and the apparent inaccuracies which stem from the liquid-gas mass transfer as discussed in Chapter 2. This complicated the use of the threshold hydrogen partial pressure below which propionate (i) is not produced and (ii) can get consumed. Sam-Soon et al., (1991b) reported these thresholds as 10^-3.7 atm and 10^-4.1 respectively. A switching function was utilised which is fully active for acetogenesis if the hydrogen concentration is 0 g/m3, 50% active when the hydrogen concentration is 1.25g/m^3 and predominately inactive above concentrations of 10g/m^3. The function used is shown in Equation (11).

$$I_{H_{2AD}} = 1 - \frac{COD[H_2]}{K_I + COD[H_2]} \tag{11}$$

where $K_I = 10$ gCOD/m^3

Figure 7 also shows the integration of the mineral precipitation due to the presence of phosphate species in the feed sludge. The inclusion of three-phase mixed weak acid-base chemistry is important for modelling AD of high P content sludge, such as WAS from nitrification-denitrification or EBPR systems (Ikumi, Harding & Ekama, 2014). The high P content of the WAS, which is released from biomass P and polyphosphate as ortho-P, affects the alkalinity of the system and can induce the formation of struvite and other mineral precipitation in the reactors (Van Rensburg et al., 2003; Harding, Ikumi & Ekama, 2011; Ikumi et al., 2013). Mineral precipitation affects the dewatering systems of the digested sludge (Van Rensburg et al., 2003). Integrating the three-phase mixed weak acid-base chemistry (including P and N) in PWM_SA_AD has the advantage to better predict pH and potential scaling problems in digested sludge dewatering systems.

While all these features are included in PWM_SA_AD (the modified version which is used in this book), several, like mineral precipitation, are not operational in this book work because modelling the methanogenic UASB system does not invoke all the included processes, in particular, the two-phase (aqueous-solid) processes involving phosphorus.

3.3 CONCLUSION

The model which is being used in this study, PWM_SA_AD, is a subset of the plant-wide PWM_SA model. This allows for plantwide models to be created with ease and without the need for transformers. The terminology used within PWM_SA was introduced here. Employing the use of standardised naming of components, results in the names being self-explanatory, thereby creating a clear understanding between modellers (because the same naming convention is used). Although some species are also components (e.g., CO_3^{2-} and PO_4^{3-}), the context in which it is used indicates whether it is a species or a component. DEs calculate the changes in components and thereafter the component concentrations are speciated into species by the effectively instantaneous AE.

One of the major advantages of PWM_SA is that the organic material composition is all specified in a way that uses the routine wastewater measurements to characterise the organic compositions. In addition, being a three-phase model, it includes mineral precipitation, allowing greater versatility to the model, especially where AD modelling of Prich WAS is required. The external speciation routine allows the precipitation and gas components to be determined in a manner which does not slow down the computational time nor cause any stiffness due to the routine operating externally from the slower bioprocesses. Because PWM_SA_AD is a subset of PWM_SA, all of the above benefits discussed for PWM_SA apply to the AD subset as well. Thus, PWM_SA_AD could be further expanded on in order to accommodate the modelling of AD dynamics. Prior to doing so, a kinetically correct and stoichiometrically mass-balanced verification was required to ensure user-confidence. This follows in Chapter 4.

4 PWM_SA MODEL MODIFICATIONS AND ADDITIONS

4.1 STOICHIOMETRY CHECKS

An essential part of model development is its reliability. This allows greater confidence in the accuracy of the model. Therefore, PWM_SA and PWM_SA_AD were subjected to rigorous checks and tests prior to calibration. Firstly, the stoichiometric equations within PWM_SA_AD were checked. This involved deriving each stoichiometric equation from first principles based on the approach of McCarty (1975). In addition to model confidence, deriving the AD equations allowed a better understanding of the stoichiometric bioprocesses, which was essential for the derivation of the sulphidogenic equations.

The approach of McCarty (1975), which was adopted by Ekama et al. (2019), entails the formulation of an electron donor equation, an electron acceptor equation (catabolism) and an equation for biomass growth (anabolism). McCarty (1975) developed a comprehensive list of half-reactions and can be used to construct most biologically mediated processes. Equation (12) is the general overall reaction for any bioprocess, as shown by McCarty (1975):

$$R = R_d - f_e R_a - f_s R_c \tag{12}$$

Where:

R = overall reaction

R_d = electron donor equation

R_a = electron acceptor equation

R_c = equation for bacterial cells (or biomass)

f_e = proportion of the electron donor which is coupled with the electron acceptor

f_s = proportion of the electron donor which is coupled with the cell formation

In order for the equation to balance, $\mathbf{f_e + f_s = 1}$. $\tag{13}$

Ekama et al. (2019) write Equation (12) in terms of the biomass yield, i.e., the number of electrons conserved in biomass in anabolism as a fraction of the number of substrate electrons utilised (the electron donor). In dynamic models which include both growth and decay of biomass, this is the actual biomass yield coefficient (Y). In steady-state models, this is the observed yield, E, which is a function of biomass yield, endogenous respiration rate and system sludge age (Ekama, 2009).

$$R = R_d - (1 - Y_x)R_a - Y_x R_c \qquad (14)$$

Where

Y is the specific organism mass yield, and x is the organism group mediating the growth.

This approach is based on COD (gO/e⁻) because the electron-donating (or accepting) capacity multiplied by 8 gO/e⁻ gives COD. This is different from the Sötemann et al. (2005b) approach, which was a mole-based approach. Shown below is the derivation of the stoichiometry for acetogenesis based on the COD method. The coefficients obtained for each component are compared with those obtained with the Sötemann et al. (2005b) approach. For the acidogenesis (both low and high pH$_2$) and acetogenesis processes, there exists a degree of freedom in the amount of H$_2$ produced and also consequently in the amount of acetate and/or propionate produced. In order to be consistent with the Mosey, (1983) approach, the amounts of products formed were kept in line with this model.

Electron Donor Reaction (R$_d$):

$$\frac{1}{14} CH_3CH_2COO^- + \frac{1}{2} H_2O \rightarrow \frac{3}{14} CO_3^{2-} + \frac{5}{14} H^+ + H^+ + e^- \qquad (15)$$

Organism Growth Equation (Y$_{AC}$R$_c$)

$$\frac{Y_{AC}k}{Y_B} CO_3^{2-} + \frac{Y_{AC}n}{Y_B} NH_4^+ + \frac{Y_{AC}p}{Y_B} PO_4^{3-} + \frac{Y_{AC}}{Y_B}(2k + 3p - n)H^+ + Y_{AC}H^+ + Y_{AC}e^- \qquad (16)$$

$$\rightarrow \frac{Y_{AC}}{Y_B} C_k H_l O_m N_n P_p + \frac{Y_{AC}}{Y_B}(3k + 4p - m)H_2O$$

Electron Destination: ((1-Y$_{AC}$)R$_a$:

$$\frac{2(1-Y_{AC})}{8+2\alpha}CO_3{}^{2-} + \frac{3(1-Y_{AC})}{8+2\alpha}H^+ + (1-Y_{AC})H^+ + (1-Y_{AC})e^-$$

$$\rightarrow \frac{1-Y_{AC}}{8+2\alpha}CH_3COO^- + \frac{4(1-Y_{AC})}{8+2\alpha}H_2O + \frac{\alpha(1-Y_{AC})}{8+2\alpha}H_2 \qquad (1$$

Overall Equation:

$$\frac{1}{14}CH_3CH_2COO^- + \left(\frac{2(1-Y_{AC})}{8+2\alpha} + \frac{Y_{AC}k}{Y_B} - \frac{3}{14}\right)CO_3{}^{2-} + \frac{Y_{AC}n}{Y_B}NH_4{}^+$$

$$+ \frac{Y_{AC}p}{Y_B}PO_4{}^{3-} + \left(\frac{3(1-Y_{AC})}{8+2\alpha} + \frac{Y_{AC}}{Y_B}(2k+3p-n) - \frac{5}{14}\right)H^+ \qquad (18)$$

$$\rightarrow \frac{Y_{AC}}{Y_B}C_kH_lO_mN_nP_p + \frac{1-Y_{AC}}{8+2\alpha}CH_3COO^- + \frac{\alpha(1-Y_{AC})}{8+2\alpha}H_2$$

$$+ \left(\frac{4(1-Y_{AC})}{8+2\alpha} + \frac{Y_{AC}}{Y_B}(3k+4p-m) - \frac{1}{2}\right)H_2O$$

Where α is the degree of freedom.

Table 8: Stoichiometry for Growth of Acetogens according to Sötemann et al. (2005b) in mol/L.

$NH_4{}^+$	CO_2	H^+	CH_3COOH	CH_3CH_2COOH	H_2	H_2O	Z_{AC}
-1	$\dfrac{1-2Y_{AC}}{Y_{AC}}$	1	$\dfrac{1-\frac{3}{2}Y_{AC}}{Y_{AC}}$	$-\dfrac{1}{Y_{AC}}$	$\dfrac{3-4Y_{AC}}{Y_{AC}}$	$\dfrac{2-5Y_{AC}}{Y_{AC}}$	1

The comparison in Table 8 below shows how that ultimately, the stoichiometric coefficients are identical. Using the yield value from Sötemann et al. (2005b) gives a different coefficient since that yield value has the units mol biomass formed/mol substrate utilised whereas the units for the yield in Equation (18) included in PWM_SA_AD is biomass COD formed/substrate COD utilised. The yield value units, therefore, have to be adjusted by multiplying by the COD per mole of biomass and dividing by the COD per mole of the substrate. In this bioprocess, the substrate is propionate. For acidogenesis, acetoclastic methanogenesis and hydrogenotrophic methanogenesis the substrates are glucose, acetate and hydrogen respectively.

In addition, the original equation from Sötemann et al. (2005b) differed in the reactants and products components selected (e.g., CO_2 for IC) compared with that used in PWM_SA (e.g., CO_3^{2-} for IC), the coefficients were adjusted to account for these differences in the following way:

- The derived Equation (18) above was first expressed in terms of 1 mole of substrate utilised, which is represented in the column "Derived Coefficients" in Table 8. Then, each coefficient was expressed in terms of 1 mole of biomass formed by dividing each coefficient by the coefficient of biomass to attain the values in the column "Derived Coefficients Adjusted".

- Sötemann et al. (2005b) used the undissociated acetate and propionate forms, whereas Equation (18) uses the dissociated forms. To account for this, the extra protons which are "given off" are added to the proton from the original Sötemann et al. (2005b) equation.

- Sötemann et al. (2005b) used CO_2 for IC instead of CO_3^{2-} as in PWM_SA. CO_2 in the (Sötemann et al. (2005b) equation was expressed as CO_3^{2-} by considering the following equation:

$$CO_2 + H_2O \leftrightarrow CO_3{}^{2-} + 2H^+$$

- Sötemann et al. (2005b) assumed the phosphorus content of the biomass to be 0, which sets the coefficient of phosphate (PO_4^{3-}) to 0 and is therefore not shown in this table.

Taking all these adjustments into account and assigning the same proportion of electrons to biomass, yields identical stoichiometric coefficients for the acetogenesis process for Sötemann et al. (2005b) and PWM_SA AD equations. All other stoichiometric reactions were checked in this way, and the appropriate adjustments to the components and yield values were made, and identical coefficients for all bioprocesses were obtained.

Table 9: Comparison of Acetogenic Coefficients

Component	Derived Coefficients (Equation (18))	Derived Coefficients adjusted[*]	Sötemann et al. (2005b) Coefficients
NH_4^+	-0.0278	-1	-1
CO_3^{2-}	0.9444	33.971	33.971
H^+	1.8749	67.443	67.443
CH_3COO^-	0.9583	34.471	34.471
$CH_3CH_2COO^-$	-1	-35.971	-35.971
H_2	2.8888	103.9137	103.914
H_2O	-2.8054	-100.914	-100.914
Biomass	0.0278	1	1

[*]Divided by 0.0278

4.2 KINETIC CHECKS

The Sötemann et al. (2005b) equations were not only derived on a mole/mole basis but also on one mole of biomass formed basis whereas the PWM_SA derived stoichiometric equations are based on one mole substrate utilised. This changes the units of the kinetic equation because Sötemann et al. (2005b) based the kinetic rate on the biomass growth rate whereas the units of the derived PWM_SA stoichiometric equation is based on the substrate utilisation rate. The biomass specific growth rate (mu or μ) is related to the substrate utilisation rate (K_m) via the yield coefficient (Y).

To compare Sötemann et al. (2005b) and the derived PWM_SA stoichiometric and kinetic rates, an Excel spreadsheet was set up wherein batch conditions were assumed for each of the 4 main bioprocesses in the anaerobic digester following hydrolysis namely acidogenesis, acetogenesis, acetoclastic methanogenesis and hydrogenotrophic methanogenesis. For each bioprocess, the substrate utilised was selected to be the substrate required for that process, i.e., glucose for testing the acidogenesis process, propionate for acetogenesis and acetate and hydrogen for acetoclastic and hydrogenotrophic methanogenesis respectively.

An initial concentration of substrate and biomass seed was selected for each bioprocess, and the Sötemann et al. (2005b) equations and constants were applied to evaluate how each concentration changed over time. The same approach was applied to the PWM_SA_AD stoichiometric and kinetic equations and the results were compared. Figure 8 below shows the substrate and biomass concentrations for the acetogenesis process for both the PWM_SA derived equations prior to the required adjustments and the Sötemann et al. (2005b) equations. As can be seen, the maximum specific growth rate is not correct for the PWM_SA derived equations. Converting the maximum specific growth rate to a substrate utilisation rate by dividing by the yield value ($\mu/Y = Km$) of Sötemann et al. (2005b) makes the predictions by the PWM_SA equations to be identical to those predicted by Sötemann et al. (2005b) approach. This illustrated the importance of ensuring that models are aligned correctly.

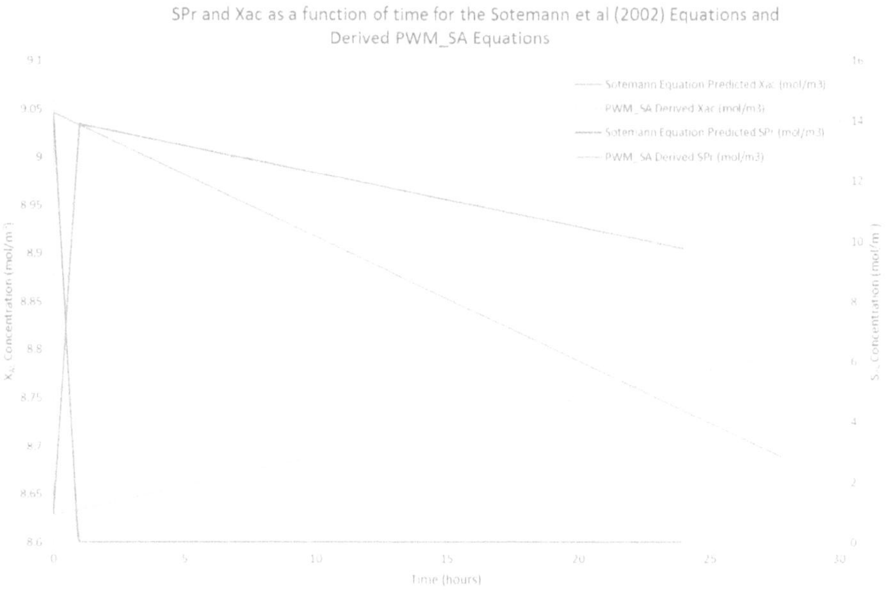

Figure 8: Graph of X_{AC} formation and S_{Pr} utilisation

4.3   MASS BALANCE VERIFICATION CHECK

Once the stoichiometric and kinetic expressions were checked and edited, mass balances could be checked stoichiometrically. In order to do this, a Microsoft Excel spreadsheet was set up based on the model stoichiometry verification method of Hauduc et al. (2010). Within this spreadsheet, the "Defined Names" function allows the names of all parameters to be defined in a similar manner to the way it is named in PWM_SA. For example, the yield of acidogens is defined as Y_AD in WEST®, and this same name can be defined in MS Excel. However, in the stoichiometric expressions in PWM_SA, the parameter is referred to as "parameters.Y_AD". To avoid lengthy expressions in the Verification spreadsheet, the "parameters." is not included and so it was removed by using the "Find and Replace" function within Excel. This minimised the chances of transcription errors occurring when typing in formulas and allows the PWM_SA model MSL code in WEST® to be directly pasted into the Excel cells with minimal adjustments.

All processes which were already part of PWM_SA_AD (Ikumi et al., 2015) were found to be 100.00% mass balanced on all the elementals (CHONPS), and COD before and after the modifications and additions were made to the stoichiometric equations. The model was also verified kinetically as part of simulation runs which will be discussed further below. This exercise provided confidence in the model allowing it to be extended for BSR applications.

4.4 SANI PROCESS ADDITIONS

Initially, it was assumed that following acidogenesis, which is a common process in both sulphidogenic and methanogenic systems, the reactions are the same in both systems, with only the COD product varying. However, it was found when analysing the stoichiometric reactions, that for sulphidogenic systems, in the degradation of propionate by the propionate-degrading SRB (or acetogenic sulphidogens), the residual sulphate (the sulphate not used in the formation of biomass) forms hydrogen sulphide which is a final COD product whereas in the methanogenic system, the propionate-degrading bacteria (or acetogens) cannot form methane directly. This in effect also means that the amount of hydrogen ions contained within hydrogen sulphide as a final COD product reduces the amount of hydrogen gas (dissolved) formed. This limited the use of methanogenic experimental systems to calibrate sulphidogenic models.

Furthermore, it was envisaged that the SANI Large Scale Plant at Sha Tin WWTP would yield reliable results for the calibration part of this study which would result in a fully calibrated SANI model capable of being used for full-scale predictions. This proposed the inclusion of multiple batch tests to determine the rate at which the influent BPO and fermentable soluble material are utilised, an important aspect for upflow reactors. However, due to numerous delays at the large-scale plant, the modelling of the SANI system aspect of this study was postponed for further investigation by other researchers once reliable results have been obtained.

The detail discussed hereunder was included since it prompted the thorough investigation of the model, which uncovered a few inconsistencies. Having resolved these inconsistencies allowed greater confidence to be placed in the model, which can then be used for further calibration.

4.4.1 SRUSB

The SANI process involves an SRUSB as discussed earlier, and the biological reactions which take place occur in an anaerobic environment. So, the existing PWM_SA_AD model can be used to model the biological reactions if they are coded into WEST®. The reactions were coded in the same way as the methanogenic equations, following the approach of McCarty (1975). Derivation and coding of the sulphidogenic reactions are what encouraged the checks on the existing PWM_SA_AD model. Similarly, in the methanogenic reactions where Sötemann et al. (2005b) derived the stoichiometry on a mole basis, Poinapen and Ekama (2010a) had done this for the sulphidogenic system. Poinapen & Ekama (2010a) had based their stoichiometric derivations on the pathways identified by Kalyuzhnyi et al. (1998). This results in the addition of three growth equations and three decay equations, assuming that butyrate and higher VFA's are not present in this system. This was shown in Appendix A. For the SANI process reactions, SO_4^{2-} is the electron acceptor which forms hydrogen sulphide. The reaction stoichiometry is given in detail in Appendix A.

In WEST®, the hydrogen sulphide component is the partially dissociated form, HS^-. The speciation routine then allocates it to the different states based on the pH, either as H_2S or HS^-. This is an important aspect since the undissociated form of sulphide is toxic to methanogenic, sulphidogenic, as well as acetogenic bacteria (Kalyuzhnyi et al., 1998).

The BSR reactions within WEST® at this time have not been coded to be interactive with the methanogenic reactions due to lack of available data but from the literature, it is evident that the sulphidogenic organisms will generally outcompete the methanogens for the available substrates. A simple switching function is used where the user can select whether sulphidogenesis or methanogenesis should be active.

Since methanogenic reactions are coded to be inactive during BSR, the effect of undissociated hydrogen sulphide on the acetogenic and methanogenic organisms is not included at this stage, but their toxicity is included in the sulphate-reducing processes. Kalyuzhnyi et al. (1998) used a linear inhibition term - shown below in Equation (19) - which Poinapen and Ekama (2010b) found to be unstable when H_2S is larger than the inhibition term, K_I.

$$I_{H_2S} = 1 - \frac{H_2S}{K_I} \qquad (19)$$

Poinapen and Ekama (2010b) thus matched the inhibition at 50% to an exponential function (see Equation (20)) which is plotted in Figure 9. This prevents the inhibition term from ever reaching zero which generally provides more stability in modelling. The K_I value differs depending on the organism group. This graph is for the K_I value of 0.185gS/L which is the inhibition constant for the acetogenic sulphidogens (propionate-degraders).

$$I = e^{-\left(\frac{[H_2S]}{0.600556K_I}\right)^2} \qquad (20)$$

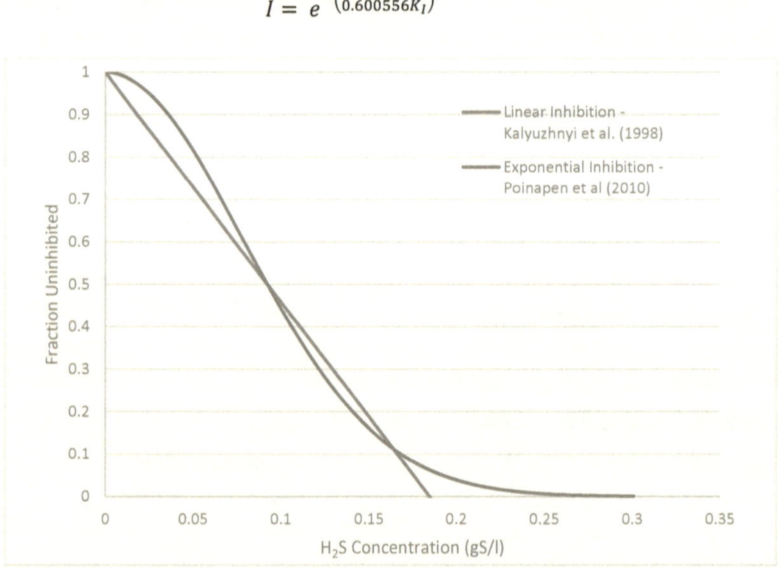

Figure 9: Difference in Linear and Exponential Sulphide Inhibition Terms

Furthermore, a sulphate switching function (shown below) is included for when sulphate concentrations are low. This turns off the BSR processes when insufficient sulphate is available to mediate the processes. The constant (K_N) was taken directly from Kalyuzhnyi et al. (1998) and differed for the different organism groups.

$$sulphate\ switching\ term = \frac{\left[SO_4{}^{2-}\right]}{K_N + \left[SO_4{}^{2-}\right]} \tag{21}$$

Since a mathematical model did not yet exist for the SANI process, the SRUSB was based on the BSR model of Poinapen and Ekama (2010b) thus for development and verification of the BSR part of the SANI process, the feed data and kinetic constants used were also from Poinapen (2008). The model was found to be 100.00% elemental mass balanced, and the predictions are reasonable with those measurements of the lab-scale BSR reactors operated by Poinapen (2008) and similar to the predictions of the model developed by Poinapen and Ekama (2010a). However, there was a large discrepancy between the measured sulphide and predicted sulphide of both models, but this was attributed to a measurement error, as stated by Poinapen, Wentzel and Ekama (2009).

Though Poinapen (2008) did not measure any gaseous CO_2, the WEST® model developed produced small quantities which may not be measurable in experimental systems. This aspect would require further calibration in future work. The model aligns with the remainder of the predictions such as large amounts of alkalinity present, high pH due to the sulphide system and low organic content in the effluent. However, the values do not match exactly. It appears that the WEST® model utilises the substrates more than the experimental systems resulting in lower effluent concentrations of substrates and sulphate. It was not possible to investigate this further since the work on the sulphidogenic system concluded at this point in order to focus on the calibration of a methanogenic model capable of predicting failure.

4.4.2 Anoxic and Aerobic Reactors

The anoxic and aerobic bioreactors at the SANI pilot plant contain suspended media which facilitate the bioprocesses as attached growth processes. These reactors are so-termed MBBR. Good nitrification and denitrification rates were observed compared with the submerged anoxic/aerobic ring-lace media filter (SAF).

A key aspect to the SANI process is low-quality electron donors that allow low sludge accumulation. The anoxic bioreactor facilitates autotrophic denitrification with HS⁻ as the electron donor. The hydrogen sulphide forms sulphate and nitrate is converted to nitrogen gas. The schematic layout was given in Figure 4. The anoxic reactor receives the effluent from the SRUSB, which is meant to be low in organic material so as to prevent heterotrophic denitrification. This further motivated the need for a well-calibrated model which can provide information as to whether any organic material will be fed into the anoxic reactor.

A recycle from the final effluent tank (which receives flow from the aerobic reactor) feeds nitrate to the anoxic reactor. Within the aerobic reactor, ammonia is converted aerobically to nitrate. Any residual hydrogen sulphide which was not oxidised to sulphate in the anoxic reactor will be oxidised in the aerobic reactor. The SANI process, therefore, requires significantly less oxygen than traditional AS systems because most (if not all) biodegradable organic material is already utilised prior to the aerobic reactor. All relevant stoichiometric derivations are included in Appendix A. These processes have also been added to the PWM_SA model in WEST®. However, they feature in the AS sub-part as opposed to PWM_SA_AD like was the case for the SRUSB.

In terms of modelling the anoxic/aerobic reactors, specific nitrification and denitrification rates were required as part of measurements which involved knowing the extent to which nitrification/denitrification occurs with the quantity of media present. This would have allowed scaling the system for large scale model predictions. Although the coding of the processes is included, and the processes are mass balanced, the nitrification and denitrification growth rates are currently not calibrated and would require additional measurements in order to be calibrated.

4.5   CONCLUSION

Following a detailed check of the model to ensure that it was kinetically and stoichiometrically accurate and mass balanced, the updated PWM_SA_AD was now ready to be used for further calibration. Although the PWM_SA_AD model is based on UCTSDM, the manner in which the stoichiometric equations are derived differ from the method used in UCTSDM. Appropriate conversion terms are therefore used to ensure that the models are equivalent. Furthermore, the importance of ensuring that the kinetic equations are based on substrate utilisation rates if the stoichiometric equations are derived in terms of the amount of substrate utilised was discussed. This ensures that that model is kinetically correct. With this, the model was ready to undergo calibration, which is presented in the next chapter.

With the additions of the SANI system processes, the model can be used to model BSR in the future, if required, or to model the full SANI system. This provides versatility to the model through the use of a simple switching function to change the use of the model from methanogenic conditions to sulphidogenic conditions. At this stage, competition between the methanogenic and sulphidogenic bacteria for the available substrate is not included due to lack of available data and scope limitations.

5 MODEL CALIBRATION

5.1 BACKGROUND

To date, many models have focused on calibration of the hydrolysis rate since it is the slowest process under most circumstances (Sötemann et al., 2005b). However, when dynamic conditions exist, or there is a build-up of inhibitory substances, hydrolysis may not be the slowest process, and thus, other processes also require calibration. Lack of calibration of the processes following hydrolysis limit the use of these models under failure conditions since in most cases, the digester failure is linked to the inhibition of the acetoclastic methanogens, thereby causing an accumulation of intermediate products. With a suitable dataset, which has an accumulation of intermediate products, calibration of the AD model may be possible.

In UASB reactors, the wastewater typically enters from the bottom of the reactor and passes upwards through the bed in a plug flow fashion (Sam-Soon et al., 1987). The plug flow feature of the UASB results in rapid consumption of substrate at the bottom of the reactor due to a high concentration of biomass. The rapid utilisation of the substrate may result in an accumulation of AD intermediate products. Figure 10 below shows the experimental set-up of the UASB operated by Sam-Soon et al. (1990). The UASB system was fed glucose at varying organic loading rates (OLR) and the concentration profiles of filtered COD (from which glucose concentration can be calculated), propionate and acetate were measured along the height of the UASB. In addition, pH, alkalinity, organic nitrogen and FSA profile measurements were also included (Sam-Soon et al., 1990; Sam-Soon, Loewenthal, et al., 1991; Sam-Soon et al., 1991b).

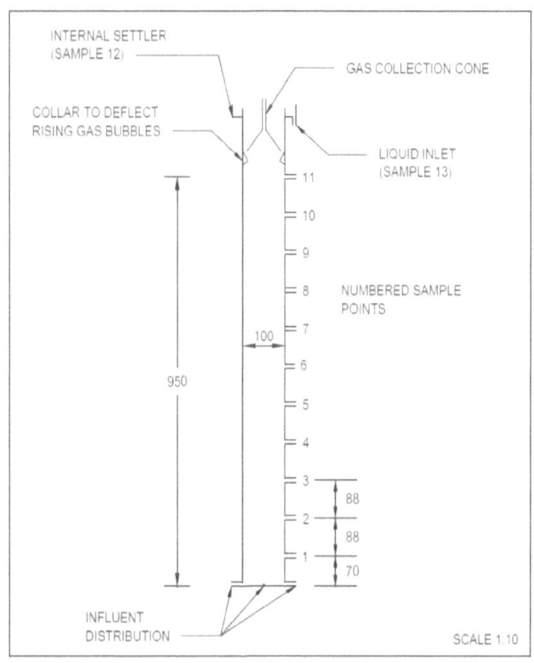

Figure 10: Schematic of the UASB Reactor Operated by Sam-Soon et al. (1990)

Generally, under steady-state conditions or plant-wide WRRF contexts, the hydrolysis process is the slowest, and the sewage sludge anaerobic digester is not subjected to strong load dynamics. Intermediate products such as glucose, acetate, propionate and hydrogen are therefore unlikely to accumulate (Ristow et al., 2005; Sötemann et al., 2005b; Ikumi et al. 2014). Due to this, earlier research focused on calibrating the hydrolysis process only. This was reasonable for AD of sewage sludge complex organics because the intermediate products are consumed as soon as they are formed under such conditions. However, under significantly changing flow and load conditions, or in high-rate systems such as UASBs, intermediate products may accumulate. For these situations, the kinetic rates of the four AD bioprocesses following hydrolysis require calibration. Therefore, because the UASBs operated by Sam-Soon et al. (1990) exhibited an accumulation of intermediate products, the configuration was modelled with PWM_SA_AD for calibration purposes. Table 10 shows the system parameters and influent characteristics.

Table 10: Influent Characteristics being fed to the UASB Reactor

Parameter	Symbol	Value	Unit
Influent Flow	Qi	15	L/day
Total COD (Glucose only)	Sti	5079	mg/L
Influent pH	pH	8.12	-
Influent Alkalinity	Alk	6000	$mgCaCO_3/L$
Reactor Volume	V	9	L
Organic Loading Rate	OLR	8.465	$kgCOD/m^3/day$
Free and Saline Ammonia	FSA	135	mgN/L

Upon entering the bottom of the UASB, the glucose is acidified at a rapid rate at the bottom of the bed and produces acetic acid and hydrogen as the products. Due to the rapid utilisation of glucose, the methanogens cannot utilise the products as fast as they are formed, resulting in an accumulation of acetate and hydrogen. The accumulation of hydrogen gas creates a high hydrogen partial pressure at the bottom of the bed resulting in the formation of propionate. The acetate and propionate accumulation results in a decline of pH. This inhibits the acetoclastic methanogens, which in turn causes further accumulation of acetate. The propionate can only be utilised once the hydrogen partial pressure has decreased to a low value again because the acetogens are inhibited under high hydrogen partial pressure. So, the hydrogenotrophs utilising hydrogen are crucial in order for the acetogens to remain uninhibited, and therefore utilise propionate. The hydrogenotrophic methanogens are reported to be sensitive to pH values below 6, which is generally not significant since, at that low pH, the acetoclastic methanogens will be completely inhibited, and thus digester failure is expected to be due to the accumulation of the acetate.

The UASB system operated by Sam-Soon et al. (1990) served as a basis for a model which is better calibrated under conditions approaching failure. Since Sam-Soon et al. (1990) fed glucose which is presumed to be utilised by the acidogens directly without requiring hydrolysis, the hydrolysis step is omitted and therefore no longer the rate-limiting step. This chapter discusses the calibration method, which was adopted as well as the challenges that arose.

5.2 UASB MODEL SET-UP

In WEST®, there were no existing UASB sub-models. As discussed above, UASBs are fed from the bottom of the reactor. The height of the reactor allows plug flow conditions to develop, which means that it can no longer be considered a completely-mixed reactor. The importance of this plug flow condition is that depending on the upflow rate applied, the biomass, which is the result of growth in the system, is able to accumulate at the bottom of the reactor and is so-termed the sludge bed. The large accumulation of the biomass results in an extremely rapid utilisation of available substrates since larger concentrations of biomass facilitate a higher bulk liquid utilisation rate.

A six in-series (ADR1 to ADR6), each completely-mixed, anaerobic digester (ADR) system was set up in WEST® to overcome the absence of an available UASB model within WEST®. The AD model allows for solids retention with a sludge retention factor (f_{ret}) in each of the in-series ADRs. The solids retention factor will enable solids which are considered "non-settleable" to pass onto the next ADR. Any solids which are "settleable" are retained in the ADR. Sludge wastage takes place at the top of the sludge bed. In the UASB model, the top of the sludge bed is at ADR5. The sixth anaerobic digester in the series (ADR6) is the supernatant above the sludge bed in which no bioprocesses take place. Six reactors were selected because it would allow modelling different experimental UASB systems which may vary in bed height. Additional reactors could have been used, but the system was seen to be very sensitive and unstable with more than six reactors.

Based on the height of the sampling ports and the diameter of the UASB, the volumes of the six in-series ADRs were determined such that samples from these ports best represent the conditions in the ADRs. The schematic of the UASB configuration is shown in Figure 11 and Table 11 gives the reactor volume for each ADR in the system.

Table 11: Sub-Reactor Heights and Volume of the WEST® integrated 6 in series UASB (WI6-UASB) and the WEST® six separate single reactor UASB (W6S-UASB).

Reactor Number	Reactor Start Height (mm)	Reactor End Height (mm)	Height (mm)	Diameter (mm)	Volume (L)
1	0	70	70	100	0.550
2	70	114	44	100	0.346
3	114	158	44	100	0.346
4	158	202	44	100	0.346
5	202	334	132	100	1.036
6	334	1146	812	100	6.377

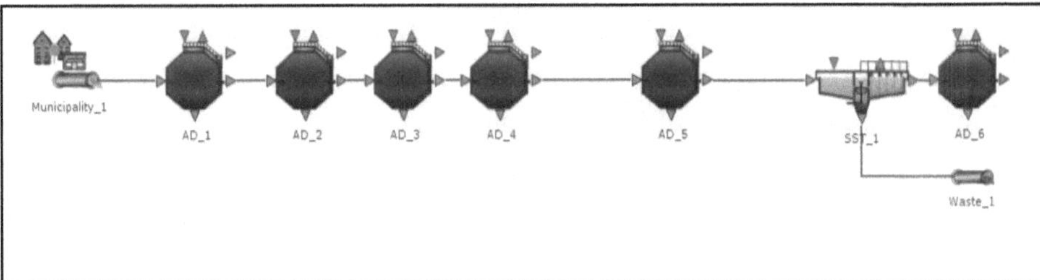

Figure 11: WI6-UASB Configuration

5.3   MODELS REQUIRED FOR THE UASB REACTOR

Model calibration was a complicated procedure due to the interaction between the:

- AD bioprocesses themselves,

- bioprocesses and aqueous phase via the pH,

- overall system SRT,

- unknown sludge distribution in the bed (the values of the retention factors), and

- possible deficiencies in the model.

The UASB system was therefore modelled in five ways in order to deal with this complexity, namely:

(1) In an Excel spreadsheet as a steady state single continuously stirred tank reactor (CSTR) AD model of volume equal to that of the total sludge bed (2.623 L) (SSSM);

(2) In WEST® as a single CSTR UASB system (WSR-UASB, Figure 12) also of volume equal to that of the sludge bed (2.623 L);

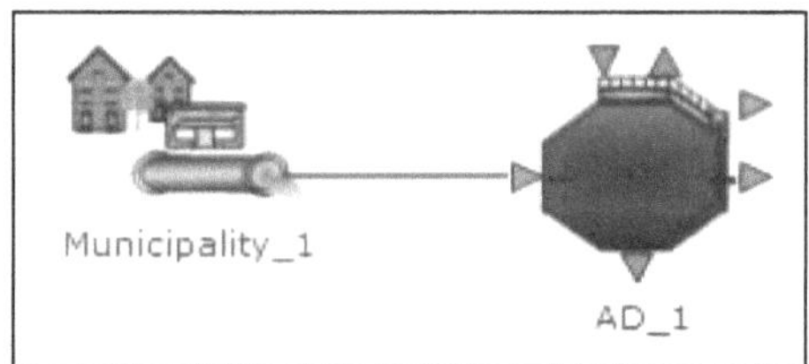

Figure 12: The UASB Modelled as a Single Completely-Mixed ADR (WSR-UASB)

(3) In an Excel spreadsheet, as a steady state five in-series CSTRs each of volume as shown in Table 11 with a combined volume equal to that of the sludge bed (2.623 L) (SS5M);

(4) In WEST®, as an integrated, six CSTRs in series UASB system (WI6-UASB) with volumes as shown in Table 11 (Figure 11); and

(5) In WEST®, as six separate single CSTRs (W6S-UASB) like Figure 12 but with each of volume as shown in Table 11 where the simulation output of the nth CSTR was set as input to the simulation of $n^{th}+1$ CSTR.

In WEST®, a steady state simulation run precedes a dynamic simulation, wherein the steady state simulation generates revised/improved starting masses of the components for the dynamic simulation. In this modelling work, the influent to the UASB was constant flow and load; therefore, only the steady state simulation feature of WEST® was used.

5.3.1 Steady State Excel Spreadsheet Models (SSSM & SS5M)

The SSSM and SS5M were set up in such a way that the overall reaction for each process is included which allows the end value of the products to be found. This is based on a steady state assumption wherein the change in concentration over time reaches zero due to the increase of a product being equal to the decrease. For example, in the case of the concentration of acetate, its' end concentration will be dependent on the amount of glucose utilised which becomes acetate, the amount of propionate utilised which becomes acetate, as well as the amount of acetate, used up to form methane.

Setting up the biological processes in a Microsoft Excel spreadsheet allows the interaction between the bioprocesses to be included, without the sensitivity of the initial mass estimates which occurs in WEST®. The calculated end results of these steady state spreadsheets (SSSM and SS5M) provided the steady-state masses, which were then used as initial masses for the biomass components in the WEST® models. This is where the value of steady-state models lies. Their results lead to substantially faster runtimes as well as superior model stability compared with assigning random values as initial masses in WEST®. So, while the end results of simulations are independent of the initial mass in the system, if the initial mass is closer to the steady-state value, the simulation reaches steady-state quicker and with less chance of "crashing" which happens when the derivative of the flux yields infinity as a solution causing the software to stop the simulation.

The steady-state AD equations of (Sötemann et al., 2005a) proved to be unsuitable for application in this study since they either (i) are derived for flow through digesters with no intentional solids retention (therefore the SRT and HRT are different), (ii) are based on biodegradable particulate COD utilisation subject to hydrolysis or (iii) assume that hydrolysis products form digester-end products and no intermediates accumulate. For this reason, the equations within SSSM & SS5M were derived from first principles. This considers the mass of product entering and exiting the system as well as the increase and the decrease in mass due to the bioprocesses. In flow-through systems, the concentration of particulate material within the digester is equal to the concentration exiting the reactor. When there is a solids retention applied, these concentrations are different.

Consider the nth reactor below in Figure 13. The concentrations of soluble material from the reactor preceding (n-1) will enter the nth reactor and will be equal to the concentration within the nth-1 reactor. This is because it won't be subjected to the effect of the solids retention factor. Similarly, the concentration of the soluble material within the nth reactor will be passed onto the next reactor. For the particulate material, because this is subject to the effect of the solids retention factor, the total solids concentration from the nth-1 reactor will not be passed onto the next, rather only the concentration multiplied by $(1-f_{retn-1})$.

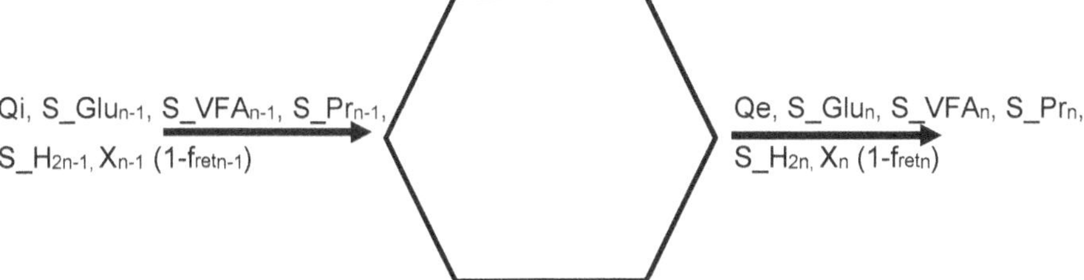

Figure 13: Steady State Nth Reactor

In such a manner then, the mass balance equation can be formulated, and because steady-state is assumed within the MS Excel spreadsheet, the derivative is set to 0. Following this, the equations can then be formulated based on solving simultaneous equations. The full derivations for SSSM and SS5M are presented in Appendix B. The same Gujer matrix approach as implemented in WEST® can be applied to solve the steady-state equations (Gujer, 2008).

The derived steady-state equations derived here are far more complex than the AD model of Sötemann et al. (2005a). This is because the equations of Sötemann et al. (2005a) were derived for an isolated reactor not receiving any inflow of biomass or intermediate products from a preceding reactor. Consider below, the mass balance equation for acidogen biomass in the case where there is no solids inflow - Equation (22) - compared with solids inflow from the preceding reactor – Equation (23). Equation (22) would be applicable to a single completely mixed anaerobic digester or the first anaerobic digester of the series because it has no biomass solids entering the digester. This is in contrast to Equation (23) wherein the amount of biomass solids entering the nth digester is set by the concentration in the nth -1 reactor and the solids retention factor of the nth-1 digester (which is f$_{retn-1}$). As can be seen, if the derivative is set to zero, in Equation (22), the substrate concentration is, in fact, independent of X$_{AD}$ but for the second case however, because X$_{ADn}$ does not appear in every term, Equation (23) is not independent of X$_{ADn}$. Therefore, there is an iterative calculation required for each biomass and associated substrate to determine the mass of each.

$$\frac{dX_{AD}}{dt} = \frac{-(1-f_{ret})}{R_h}X_{AD} + \frac{\mu_{AD}\frac{C_{Gl}}{MW_{Glu}}}{Ks_{AD} + \frac{C_{Gl}}{MW_{Glu}}}X_{AD} - b_{AD}X_{AD} \tag{22}$$

$$\frac{dX_{AD}}{dt} = \frac{(1 - f_{ret\,n-1})}{R_h} X_{AD_{n-1}} - \frac{(1 - f_{ret\,n})}{R_h} X_{AD_n} + \frac{\mu_{AD} \frac{C_{Gl}}{MW_{Glu}}}{Ks_{AD} + \frac{C_{Gl}}{MW_{Glu}}} X_{AD_n} - b_{AD} X_{AD_n} \qquad (23)$$

Because Sam-Soon et al. (1990) did not provide the sludge age, nor solids concentrations within the bed for their UASB systems, the f_{ret} could only be determined by finding the best fit to their measured bed concentrations of glucose, acetate and propionate and then using the determined f_{ret} (or $f_{ret\,n}$ depending on if it is the first digester in the series or not) to find the SRT based on the relationship shown in Equation (24) below. From the equations within SSSM and SS5M, the initial biomass masses for the WEST® models could be determined using Equation (24).

$$f_{ret} = 1 - \frac{V_{liq}}{SRT \times Q_{out}} \qquad (24)$$

The mass of water required for the start of the simulation was simply the volume of water converted into grams. The initial masses of other components could be set to any low non-zero value to prevent computational errors caused by setting initial masses to zero.

So, in the case of testing the default kinetic constants, the kinetic constants would be held at their default values as given by Sötemann et al. (2005b). The retention factors for each in-series reactor could then be varied in order to attempt to match the glucose concentration. This resulted in iterations required between the (i) glucose concentration and acidogen concentration, (ii) propionate and acetogen concentration with assumed hydrogen inhibition terms and (iii) acetate and acetoclastic methanogen concentration with assumed pH and hydrogen inhibition terms. The hydrogen concentration and hydrogenotrophic methanogen concentration were also found by iteration, which was used to determine the actual hydrogen inhibition terms. Following this, an iteration is required in order to get the actual hydrogen inhibition terms to match the assumed hydrogen inhibition terms. The concentrations of acetate, propionate and hydrogen with their associated biomasses are then recalculated in a loop wise manner until the results converge. The concentrations of the biomasses were then converted into mass by multiplying by the reactor volume. This mass was entered as the initial reactor mass in WEST® and gave a better estimate thereby promoting greater model stability. All new experimental system runs in WEST® for which good initial biomass masses were not available from previous runs were started this way.

5.4 INITIAL CALCULATIONS

Prior to any accurate model predictions, the reactor configurations need to be set up in WEST®. This allows the user to build on and make the modelling environment more complex in a step-wise manner, which reduces the chances of errors and extensive debugging. As a starting point, therefore, the influent file was first characterised.

The influent file is based on what was fed to the UASB by Sam-Soon et al. (1990). However, it needed to be made compatible with the components which WEST® requires (see Table 5). This means, therefore that glucose needed to be expressed in terms of actual mass of glucose as opposed to COD concentration. This also applied to all other components such as ammonia and phosphate. They would need to be of the form NH_4^+ and PO_4^{3-} respectively.

The overall approach to characterizing the influent aqueous phase to align with the model components is to measure (i) the ionic strength via electrical conductivity (EC), (ii) the weak acid/bases that affect the pH such as H_2CO_3 alkalinity and pH for the IC system, FSA, OP and acetate (and propionate) via the 5-point titration of Moosbrugger et al. (1993d) and (iii) the main cations involved in the reactions of interest such as Mg, Ca, Fe^{2+} and Fe^{3+}. The ionic strength of all the species of the weak acid/bases is lower than that measured by the EC. So, Na^+ and Cl^- are added to (i) match the calculated ionic strength to that measured via the EC and (ii) equalise the positive (+ve) charges with the negative (-ve) charges. This approach ensures that both dissociation constants and ion-pairing stability constants are correctly adjusted for ionic strength.

In order to convert the weak acid/base species into the form required for WEST®, the equilibrium dissociations are used. However, dissociation constants are usually reported at 25°C with ideal conditions and infinite dilution. A table of dissociation constants for the most common weak acids/bases found in wastewater is given by Loewenthal, Ekama and Marais (1989) with their temperature dependence included.

Tait et al. (2012) discuss the impact of assuming ideal (infinite dilution) conditions when such conditions do not apply. For different waters (sludge, effluent reverse osmosis concentrate, cow slurry and sea water), the measured pH was compared against simulated pH which either had no corrections to ionic activity, ionic activity corrections only or ionic activity corrections with ion pairing. Depending on the ionic strength of the water, the simulated pH without any correction predicts a pH which is higher than the measured pH. If a model does not make appropriate corrections, the model-predicted pH could be higher than the actual pH, which could mask any instability in the system.

The corrections for nonideal conditions include ionic strength adjustment of pK values and inclusion of ion pairs if required. If the ionic strength is less than 0.08 mol/kg, then ideal conditions can be assumed without any corrections to pK values. For ionic strengths between 0.08 and 0.2 mol/kg, corrections to the ion activity need to be included which changes the equilibrium state of the dissociation and association reactions. Inclusion of ion pairs, as well as correction to the ionic activity, is required if the ionic strength is greater than 0.2 mol/kg (Tait et al., 2012)

The speciation routine used in WEST® (Brouckaert, Brouckaert & Ekama, 2019b) includes both ionic activity correction and ion pairing. The ion pairs have the effect of removing ions from the aqueous phase thereby further reducing the "felt pressure" of the charge of the ion in the aqueous phase. As the ionic strength decreases, the ions "absorbed" by the ion pairs are returned to the aqueous phase.

For influent feed solutions, the ionic strength is generally below 0.2 mol/kg unless there is a seawater component. Thus, the pre-processor spreadsheet included ionic strength corrections to the activity coefficients, but not ion-pairing. Although ion pairing is included in the ionic speciation routine, negligible ion pairs are formed for low ionic strength wastewater.

In cases where conductivity is measured, it can be translated to a total ionic strength based on Equation (25) shown below from (Bhuiyan, Mavinic & Beckie, 2009).

$$I = 7.22 \times 10^{-6} EC_{25} \tag{25}$$

With the temperature dependency given by:

$$EC_{25} = \frac{EC_T}{1 + a(T - 25)} \tag{26}$$

where $a = 0.0198°C^{-1}$

The ionic strength based on the conductivity, represents the total ionic strength of the solution. This ionic strength is used to determine the activity coefficients. The Davies Equation – Equation (27) - is useful to calculate the activity coefficients if the ionic strength is less than 0.5 mol/kg and has been found to be more accurate than the Debye-Hückle equation and extended Debye-Hückle equation for ionic strengths higher than 0.005 and 0.01 mol/kg respectively (Tait et al., 2012).

$$log f_i = -AZ_i^2 \times \left(\frac{\sqrt{I}}{1+\sqrt{I}} - 0.3I\right) \tag{27}$$

With the temperature dependence being $A = \frac{1.825 \times 10^6}{78.3T^{1.5}}$ where T is in Kelvin

Based on Equation (27) above, the monovalent, divalent and trivalent ion activity coefficients can be determined. These coefficients are included in the kinetic dissociation reactions, as shown below:

$$K_{eq} = \frac{f_c[C]^c f_d[D]^d}{f_a[A]^a f_b[B]^b} \tag{28}$$

Based on the equilibrium reaction: $A^a + B^b \leftrightarrow C^c + D^d$

where f_a, f_b, f_c and f_d are the ion activity coefficients.

The square brackets refer to molal concentrations (mol/kg solvent). Activity coefficients multiplied by molal concentrations results in activity. Taking negative logs of the above expression results in an equation for pK, which is adjusted for ionic activity, called pK'.

Once the dissociation constants have been corrected for temperature and ionic strength, they are used to speciate the components into their actual species concentrations based on the pH. The species are then used to determine the ionic strength (from the known species) based on Equation (29), and this can be compared with the ionic strength of the known species:

$$I = 0.5 \times \sum_{i=1}^{n} S_{[i]}Z_i^2 \tag{29}$$

where

$S_{[i]}$ is the species concentration

Z_i is the charge on the species

The difference between the ionic strength calculated from the conductivity (actual) and the ionic strength calculated from the equations above (from known measured species) is made up by adding Na^+ and Cl^- concentrations to represent the non-measured cations and anions such that (i) the ionic strength of the +ve charges equals that of the -ve charges and (ii) the ionic strength of the +ve and -ve charges equal that measured by means of conductivity. The Na^+ and Cl^- concentrations added to form part of the WEST® influent file. As a final step, the species are converted into WEST® components in the form required.

The above procedure of equalising the measured and modelled ionic strength and establishing the charge balance requires iterations because the pK values change with ionic strength. If the ionic strength of measured anions and cations are IS_{an} and IS_{cat} and the added sodium ions and chloride ions is IS_{Na} and IS_{Cl} respectively, then the ionic strength from the measured EC, IS_{meas} is given by:

$$IS_{meas} = IS_{an} + IS_{cat} + IS_{Na} + IS_{Cl}$$

and

$$\frac{1}{2} IS_{meas} = IS_{an} + IS_{Cl} = IS_{cat} + IS_{Na}$$

from which

$$IS_{Na} = \frac{1}{2} IS_{meas} - IS_{cat}$$

and

$$IS_{Na} = \frac{1}{2} [Na^+]$$

so

$$[Na^+] = IS_{meas} - 2IS_{cat}$$

Also,

$$IS_{Cl} = \frac{1}{2} IS_{meas} - IS_{an}$$

and

$$IS_{Cl} = \frac{1}{2} [Cl^-]$$

so

$$[Cl^-] = IS_{meas} - 2IS_{an}$$

Where the ionic strength calculated from the measured EC is IS_{meas}

and IS_{cat} and IS_{an} are the ionic strengths calculated from the known positive and negative charged species, respectively. The iteration converges quickly because the pK values do not change much with changes in S_Na and S_Cl added. The iteration continues until a charge balance is achieved.

When actual sewage or primary sludge is not the influent (as was the case with the dataset of Sam-Soon et al. (1990)), but is instead made up synthetically in a laboratory, the pre-processing calculations differ because conductivity may not have been measured. In this case, the ionic strength needs to be calculated iteratively by initially assuming an estimated ionic strength. The dissociation constants are then computed using the estimated ionic strength value, which allows the species concentrations to be determined. From this, an ionic strength can be calculated, and an error will be found. A loop or solver function in Microsoft Excel (Visual Basic) is then used to remove the error by setting the error to 0 by changing the estimated ionic strength value. Once the correct ionic strength is determined, the species concentrations are calculated and then converted into the form required by WEST®.

As shown in the Components list, the WEST® components are selected to align with other speciation models such as MINTEQA2 (Allison, Brown & Novo-Gradac, 1991) or PHREEQC (Parkhurst & Appelo, 2013). This means, therefore, that the BA fed would have to be aggregated into a total carbonate species and then based on pH, fractionated into the H_2CO_3, HCO_3^- and CO_3^{2-} species depending on the pK values and pH as described above. Following this, the aqueous phase is "de-speciated" to determine the component concentrations. For example, hydrogen ions are "removed" from the HCO_3^- and H_2CO_3 species and added to H^+ component (S_H) and the remaining CO_3^{2-} added to the CO_3^{2-} component (S_CO3) to obtain the concentration of the components. This is done for each weak acid/base system present in order to obtain the influent component concentrations in terms of the model selected components inter alia CO_3^{2-}, PO_4^{3-}, HS^-, NH_4^+, Ac^-.

As given in Table 10, Sam-Soon et al. (1990) fed glucose as the substrate which has the advantage of an exactly known composition as opposed to the apple juice which was fed in the first UASB system experiments (Sam-Soon et al., 1987). In addition to the glucose, Sam-Soon et al. (1990) also fed a trace element and nutrient solution which was given by Sam-Soon et al. (1987) as well as excess ammonia. The full list of trace element and nutrient solution, shown in Table 12 and Table 13 below respectively, were obtained from by Sam-Soon et al. (1987).

Table 12: Trace Element Solution

Trace Element Solution	
Compound	g/L
H_3BO_3	0.05
$FeCl_2$	2
$ZnCl_2$	0.05
$MnSO_4$	0.5
$CuCl_2.2H_2O$	0.03
$(NH_4)_6Mo_7O_{24}.4H_2O$	0.05
$AlCl_3.6H_2O$	0.05
$CoCl_2.6H_2O$	2
$MnCl_2$	0.25
$MgCl_2$	1
EDTA	0.05
KI	0.05
$NiCl_2.6H_2O$	0.25
HCl (conc)	1mL

This trace solution required speciation prior to being used in WEST®. The amount of trace solution fed was 3mL/L of influent. In order to speciate this trace element solution, and hence determine the ionic strength, the concentration of the compounds given by Sam-Soon et al. (1987) was converted to mol/L by dividing by the molar mass of that compound. This was then multiplied by the amount of feed (3mL/L) in order to determine the actual concentration of the compound in the feed. In order to determine the concentrations of the ions, some assumptions were required:

- Firstly, metal salts were assumed to dissociate completely into the respective ions which they comprise.

- No information regarding the dissociation of ammonium molybdate tetrahydrate could be sourced. It was therefore assumed that it dissociates into ammonium ion, heptamolybdate ion and water.

- Three dissociation constants were used for Boric acid dissociation with the products being H_3BO_3, $H_2BO_3^-$, HBO_3^{2-} and BO_3^{3-}.

- Four dissociation constants are found in the literature for ethylenediaminetetraacetic acid (EDTA) dissociation with the abbreviated products being H_4Y, H_3Y^-, H_2Y^{2-}, HY^{3-} and Y^{4-}. However, due to the minuscule amount of EDTA added, and specifically if the pH is known to be at 8.12 from the system operated by Sam-Soon et al. (1990), almost 100% of EDTA will be of the form HY^{3-} according to the graph provided by Skoog et al. (2004). It, therefore, seemed reasonable to assume that all EDTA is of the form HY^{3-}.

The charged ions were then used to determine the total ionic strength of the trace element solution. Because many species of the trace solution are not components in WEST®, S_Na and S_Cl were used in order to match the total ionic strength of the trace element solution, based on the procedure for IS_{meas} discussed previously. Lastly, the mol/L concentrations were multiplied by the molar mass of each component and added to the WEST® influent file because WEST® uses mg/L for its components.

The nutrient solution which was fed by Sam-Soon et al. (1987) is shown below and Table 13. A minimum quantity of nutrient solution required according to Sam-Soon et al. (1987) is 50mg NH_4Cl and 10mg K_2HPO_4 per 1000mg COD. However, excess ammonia was also fed to avoid limitations on biomass growth and granule formation. The concentration of FSA was available from the influent experimental results while the nutrient solution details below allow the concentrations of potassium (K), chloride (Cl) and phosphate species to be calculated. The influent ammonia was speciated into ammonia and ammonium ions based on the pH, chloride ions were added as S_Cl, and similarly for potassium ions. The phosphate concentration was speciated into PO_4^{3-}, HPO_4^{2-}, $H_2PO_4^-$ and H_3PO_4. This speciation was necessary to correctly de-speciate the phosphate into S_H and S_PO4 components as explained above for the IC system. This same approach applies for the ammonia system which in WEST® is represented by the NH_4^+ ion. So, for any ammonia (NH_3), H^+ is removed from the S_H component in the influent in order to be expressed as NH_4^+. Following the full fractionation and pre-processor step, the influent file is generated into a text file and read by WEST®.

Table 13: Nutrient Solution

Nutrient Solution	
Compound	g/L
NH_4Cl	5.00
K_2HPO_4	2.00

5.5 IDENTIFYING IMPORTANT PARAMETERS

(The more sophisticated statistical methods employed hereunder such as the lasso method were completed with the assistance of Mr Hassan Sadiq who is a PhD candidate in the Statistical Sciences Department at the University of Cape Town. Further detail of the statistical method is available in Appendix C)

5.5.1 Background

Model calibration typically involves either tuning certain parameters manually or by means of a parameter estimation exercise. Other parameters are held fixed, but this can lead to the calibrated parameters being biased to the fixed values. This problem exists due to the interaction between parameters or the overlooking of parameters which may not have been identified as important. Overcoming this problem by following a protocol which includes sensitivity analyses proved to be a valuable tool (Brun et al., 2002). The BIOMATH protocol (Vanrolleghem et al., 2003) indicates the important steps required in model calibration, which includes an iterative procedure between sensitivity analysis and calibration by means of parameter estimation. A calibration methodology was proposed by Brun et al. (2002) which was adapted to this study as best as possible.

Brun et al. (2002) discusses that once the initial experimental system has been set up (discussed in Section 5.3 above), the associated uncertainty within parameters needs to be identified. To a large extent, this is based on experience and literature values. Brun et al. (2002) highlighted typical expected uncertainty for certain classes. For example, the composition of biomass is well known and does not differ widely from the common $C_5H_7O_2N$. Similarly, in AS systems, the yield of the OHOs is normally very close to 0.45gVSS/gCOD. However, in AD systems, the yield values may not be as close to a constant, but literature values support narrow ranges of uncertainty. Table 14 below indicates the range for each of the parameters in the sensitivity analysis.

Various methods exist to identify important parameters (called factors for the statistics area), non-influential factors and for some methods, the interaction between factors. Literature sources compare the results of a few methods namely standard regression coefficient (SRC) method which is based on fitting a linear model to the dataset, Morris Screening (Morris, 1991), which is a one-at-a-time approach, and Extended-FAST (Fourier Amplitude Sensitivity Testing) (Neumann, 2012; Cosenza et al., 2013), which follows an analysis of variance (ANOVA)-like calculation of sensitivity indices (Saltelli, Tarantola & Chan, 1999). It was found in the study conducted by Neumann (2012) that although SRC was applied outside its' validity range, it still identified similar important parameters to Extended-FAST while the Morris Screening did not identify as many. However, Morris Screening tends to focus primarily on factor fixing rather than factor prioritisation. The SRC method is useful because it assumes that a linear relationship exists between model factors and model outputs. Although this is not necessarily true for biological systems, if the correlation coefficient is greater than 0.7, it can be assumed that the applied linear model explains the relationships reasonably well and accounts for 70% of the variance in the data. Consequently, information about the most significant parameters can be accepted to be true. Linear models are far simpler to understand compared with the more complex analysis such as Extended-FAST, which is ANOVA based, especially for users who are not familiar with detailed statistical methods. This was the reason for selecting the SRC method as the basis of this sensitivity analysis.

5.5.2 Preliminary Analysis

For the purposes of the statistical analysis discussed hereunder, the parameters in PWM_SA which are used in this analysis (extracted from Table 6) are referred to as "primary explanatories", and the variables (extracted from Table 5) **contained** within the linear models are termed "secondary explanatories". This is because, in statistical jargon, any measurement collected over multiple independent replicates is referred to as a *variable* and therefore, the renaming of parameters and variables to primary and secondary explanatories respectively is to avoid confusion with the WEST® modelling software use of the term.

The actual variable for which a linear model is being derived is termed "output variable". So, S_Pr, which is a component in PWM_SA_AD (in Table 5) will be a secondary explanatory for all linear models (in which it is included as a term) excluding the linear model for itself where it is an output variable instead. Consider Equation (30). In this case, S_Pr is the output variable. S_H2 and S_Glu are secondary explanatories, and f_{ret} is a primary explanatory. However, if instead, Equation (31) is considered, S_H2 is now the output variable with S_Pr and S_Glu being the secondary explanatories and f_{ret} is the primary explanatory.

$$S_{Pr} = \beta_0 + \beta_1 S_{H2} + \beta_2 f_{X_{Ret}} + \beta_3 S_{Glu} \tag{30}$$

$$S_{H2} = \beta_0 + \beta_1 S_{Pr} + \beta_2 f_{X_{Ret}} + \beta_3 S_{Glu} \tag{31}$$

Uncertainty analysis was performed with the standard Monte Carlo method. The uncertainty ranges were either determined from literature (Brun et al., 2002), expert knowledge or investigations using the steady state models and allowed the upper and lower bounds to be identified based on the default values. The upper and lower bounds are listed below in Table 14 for the primary explanatories used in the sensitivity analysis. This complete list of primary explanatories was tested to prove that many are non-significant.

Table 14: Uncertainty Ranges for the Primary Explanatories within PWM_SA_AD

Parameter Name	Default Value	Lower Bound	Upper Bound
ISS_BM	0.15	0.149625	0.1575
KS_AC	0.089	0.06675	0.1335
KS_AD	0.781	0.58575	1.1715
KS_AM	0.013	0.00975	0.0195
KS_BInf_AD_hyd	10.124	7.593	15.186
KS_BOrg_AD_hyd	10.37	7.7775	15.555
KS_HM	0.156	0.117	0.234
K_I_H2	9.999375	7.499531	14.9990625
K_I_H_AD	0.0155	0.011625	0.02325
K_I_H_AM	1.15E-06	8.63E-07	0.000001725
K_I_H_HM	0.00053	0.000398	0.000795

Parameter Name	Default Value	Lower Bound	Upper Bound
K_I_H_Meth	0.0001	0.000075	0.00015
K_I_NH3	0.0018	0.00135	0.0027
Y_AC	0.039714286	0.039615	0.0417
Y_AD	0.0895	0.089276	0.093975
Y_AH	0.0895	0.089276	0.093975
Y_AM	0.03925	0.039152	0.0412125
Y_HM	0.04	0.0399	0.042
b_AC	0.015	0.0144	0.018
b_AD	0.041	0.03936	0.0492
b_AM	0.037	0.03552	0.0444
b_HM	0.01	0.0096	0.012
f_XU_Bio_lysis	0.08	0.0798	0.084
f_{ret}	0.1	0.999	0.5
i_Ca_PP_mol_perP	0.053	0.03975	0.0795
i_H_Org_mol_perC	1.4	1.3965	1.47
i_HF_mol_perC	2.0884684	2.00493	2.50616208
i_HU_mol_perC	1.646	1.58016	1.9752
i_H_XBInf_mol_perC	2.0884684	2.00493	2.50616208
i_H_XBOrg_mol_perC	1.4	1.3965	1.47
i_H_XUInf_mol_perC	1.4	1.3965	1.47
i_H_XUOrg_mol_perC	1.4	1.3965	1.47
i_K_PP_mol_perP	0.312	0.234	0.468
i_Mg_PP_mol_perP	0.297	0.22275	0.4455
i_N_Org_mol_perC	0.2	0.1995	0.21
i_NF_mol_perC	0.13548408	0.130065	0.162580896
i_NU_mol_perC	0.062	0.05952	0.0744
i_N_XBInf_mol_perC	0.13548408	0.130065	0.162580896
i_N_XBOrg_mol_perC	0.2	0.1995	0.21
i_N_XUInf_mol_perC	0.2	0.1995	0.21
i_N_XUOrg_mol_perC	0.2	0.1995	0.21
i_O_Org_mol_perC	0.4	0.399	0.42
i_OF_mol_perC	0.43266815	0.415361	0.51920178
i_OU_mol_perC	0.593	0.56928	0.7116
i_O_XBInf_mol_perC	0.43266815	0.415361	0.51920178
i_O_XBOrg_mol_perC	0.4	0.399	0.42
i_O_XUInf_mol_perC	0.4	0.399	0.42
i_O_XUOrg_mol_perC	0.4	0.399	0.42
kH_F_AD_hyd	10	7.5	15
kH_PP_AD_hyd	1	0.75	1.5
kM_BInf_AD_hyd	2.004	0.72144	3.6072
kM_BOrg_AD_hyd	1.95	0.702	3.51
kM_fPP_PAO_PHAstor	0.3	0.225	0.45
mu_AC	1.15	1.104	1.38

Parameter Name	Default Value	Lower Bound	Upper Bound
mu_AD	0.8	0.6	1.2
mu_AM	4.39	1.5804	7.902
mu_HM	1.2	0.432	2.16

Following an initial linear model fit with 1000 runs, for each variable for which measured data were available from Sam-Soon et al. (1990), the statistically significant primary explanatories were retained for further analysis. These are identified by considering the p-values. Consider a linear equation with no interaction as in Equation (32). The β values are the coefficients of the primary explanatories x_1, x_2 used in a linear model to obtain a good fit (to the WEST® results) for the output variable (y). β_0 is the intercept. The p-values provide an indication of the likelihood that the β values are different from the values displayed. Therefore, a low p-value means that the β values are statistically relevant since there is a low chance that they are incorrect. The revised primary explanatory list is given in Table 15.

$$y = \beta_0 + \beta_1 x_1 + \beta_2 x_2 \ldots \tag{32}$$

Table 15: Revised Primary Explanatory List for Sensitivity Analysis

Parameter Name	Default Value	Lower Bound	Upper Bound
KS_AC	0.09	0.06675	0.1335
KS_AD	0.78	0.58575	1.1715
KS_AM	0.01	0.00975	0.0195
KS_HM	0.156	0.117	0.234
K_I_H2	9.999375	7.499531	14.9990625
K_I_H_AD	0.0155	0.011625	0.02325
Y_AC	0.039714286	0.039615	0.0417
Y_AD	0.0895	0.089276	0.093975
Y_AH	0.0895	0.089276	0.093975
Y_AM	0.03925	0.039152	0.0412125
Y_HM	0.04	0.0399	0.042
b_AC	0.015	0.0144	0.018
b_AD	0.041	0.03936	0.0492
b_AM	0.037	0.03552	0.0444
b_HM	0.01	0.0096	0.012
f_{ret}	0.1	0.999	0.5
i_H_Org_mol_perC	1.4	1.3965	1.47
i_N_Org_mol_perC	0.2	0.1995	0.21
i_O_Org_mol_perC	0.4	0.399	0.42
mu_AC	1.15	1.104	1.38

Parameter Name	Default Value	Lower Bound	Upper Bound
mu_AD	0.8	0.6	1.2
mu_AM	4.39	1.5804	7.902
mu_HM	1.2	0.432	2.16

The primary explanatories which were deemed intuitively to be insignificant were also found to be statistically irrelevant - the system of Sam-Soon et al. (1990) was not a very complex system with multiple organic types. Therefore, the primary explanatories which were removed was based on both intuition and statistical analysis. For a few output variables, the linear model resulted in a regression coefficient which was not greater than the literature threshold of 0.7 which indicates that not a large amount of variance can be explained with the linear model. These variables required further investigation, which is discussed below. At this stage, it was not clear why the obtained coefficient of determination (R^2) was not greater than 0.7. It could be due to (i) the lack of interaction terms, (ii) lack of inclusion of secondary explanatories (variables within PWM_SA_AD) or (iii) lack of interaction between the primary and secondary explanatories. Because case (iii) is the "worst" case scenario which requires the greatest number of simulation runs in order to generate a large dataset, additional simulation runs were completed to cater for this scenario.

5.5.3 Detailed SRC Analysis

5.5.3.1 Initial Analysis

For the WSR-UASB system, there were output variables which met the criteria for the regression coefficient greater than 0.7 as an indication that the SRC method (linear model) without interactions is a valid means of sensitivity. However, other output variables which did not have a high correlation coefficient needed to be investigated further to identify why this was the case. For example, the propionate concentration (S_Pr) does not only depend on the value of the so-called primary explanatories, but also the secondary explanatories, glucose (S_Glu) and the hydrogen inhibition function dictated by the hydrogen concentration. This, therefore, meant that all secondary explanatories which could contribute to the output variable concentration needed to be included in the multi-variate SRC analyses. Interactive terms may also play a significant role in obtaining a high correlation coefficient. Equation (33) below shows a linear equation with interactive terms. So, the effect of x_1 on y now depends not only on β_1 but also on β_3 and x_2 and similarly for x_2. This changes the β values compared with Equation (32). Possible interactions also need to be included as part of the investigation in case it is deemed significant.

$$y = \beta_0 + \beta_1 x_1 + \beta_2 x_2 + \beta_3 x_1 x_2 \ldots \tag{33}$$

To avoid bias, where interactions are required, all possible interactions were included, which results in a significant number of simulation runs required. Because secondary explanatories inclusion was also expected, this increases the required amount of runs even further. The new dataset comprised 13104 observations (model runs) with 23 primary explanatories and 18 secondary explanatories. If one considers the number of possible pairwise interaction terms in the entire dataset (primary and secondary explanatories) for an output variable then mathematically, it would be $\binom{40}{2}$ (40 choose 2) resulting in 820 pairwise terms. The dataset needs to be in far excess of that in order to obtain meaningful results, which is the reason for the large dataset in excess of 13 000 runs.

For the WSR-UASB, the primary and secondary explanatories were subjected to an initial correlation analysis by considering the correlation between two primary explanatories, two secondary explanatories or a primary and a secondary explanatory while not considering the others. This provided an idea of how the data is related — for example, the correlation between KS_AC with each of the secondary explanatories such as Alkalinity and S_Glu. The correlation between the primary explanatories was very close to 0 as expected because the primary explanatories are not related to each other in any way and merely randomly generated in WEST® as part of the simulation results. The correlation was also checked between secondary explanatories such as Alkalinity and H_2CO_3 Alkalinity. Figure 14 contains a graphical illustration of the pairwise linear relationships between them.

There are some interesting features in the bottom-left rectangle that contain the pairwise relationships between the primary and secondary explanatories. The relationship between f_{ret} and the secondary explanatories is the most pronounced among all its primary counterparts. However, mu_AD, mu_AM and mu_HM also appear to have a slightly significant relationship with the secondary explanatories. Given these apparent relationships between primary and secondary explanatory pairs, f_{ret}, mu_AD, mu_AM and mu_HM are expected to be prominent determinants of the secondary explanatories.

One would expect certain output variables to be highly correlated such as Alkalinity and H_2CO_3 Alkalinity because Alkalinity has within its expression H_2CO_3 Alkalinity. This was also the case between TSS and VSS which is due to TSS and VSS being related through ISS, and S_NH and FSA with FSA simply being S_NH expressed in terms of mgN/L. Strongly correlated variables bring the same information to the statistical model and may cause difficulty in solving the linear model. It was for this reason that VSS, FSA and H_2CO_3 Alkalinity were removed from further analysis.

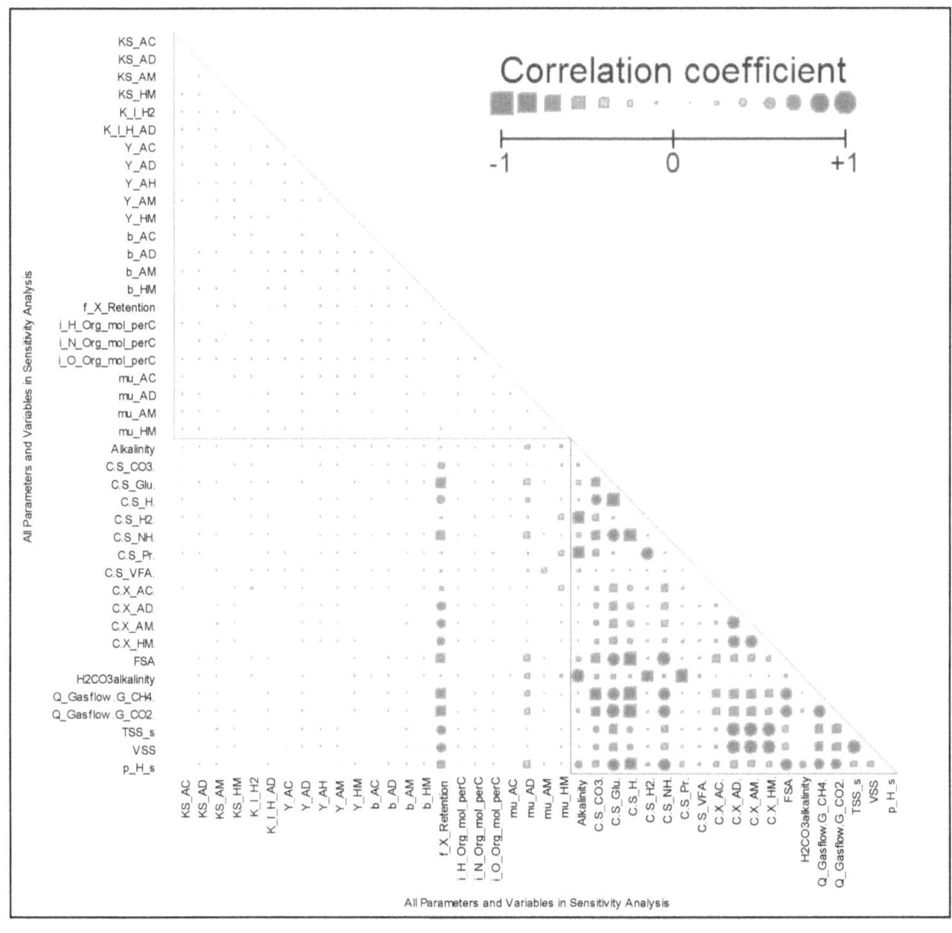

Figure 14: Correlation Coefficients between 2 parameters or variables

At this stage, the data was standardised by means of subtracting the mean and dividing by the standard deviation. Standardised values allow comparison between the coefficients for each explanatory, but in certain cases, unintended loss of information could result with the use of standardised values.

Following this, a box plot of the uncertainty ranges for each primary and secondary explanatory was plotted. Full details can be obtained in Appendix C. The box plots aim to show how the data is distributed and identify any outliers if applicable. Of particular interest was run number 9466. In this case, the value of S_VFA was more than 30 times the standard deviation of this explanatory. Furthermore, run numbers 1007, 1524, 2153 and 5623 had a result of 0 for each secondary explanatory. Upon further analyses, it was found that these (the results of 0) were the result of an incomplete simulation run which occurs as a result of a mathematical instability in the system brought about by initial masses which are too far off from the steady-state values. Because these experimental results were generated in a large overall simulation run, one is not able to set a different initial mass for each run. However, because it was only four observations in which this occurred, this aspect was not considered to be problematic. These five observations (1007, 1524, 2153, 5623 and 9466) were removed in the remaining analyses.

5.5.3.2 Analysis with only Primary Explanatories - No Interaction

The results with 13 099 observations were analysed in R (R Core Team, 2019) by assuming linear relationships without interactions. The variables listed in Table 16 below are those for which high correlation coefficients (>0.7) were obtained. These indicate that the simplest linear model can explain the dataset reasonably well. The highest beta coefficients would be the most important explanatories because the dataset has been standardised.

Table 16: Output Variables which have an Adjusted R^2 greater than 0.7

Variable	Adjusted R^2
S_Glu	0.73175
Q_GasflowCH4	0.73078
Q_GasflowCO2	0.73645

5.5.3.3 Analysis with only Primary Explanatories Including Interactions

For the output variables which did not have a high correlation coefficient, the analysis was repeated to include interactions between primary explanatories. Interactions become important when the magnitude of one explanatory influences the magnitude of a second explanatory. A very simple example is considering the preference of condiments on a sampled population between chocolate sauce and tomato sauce. But the condiment preference depends on the food on which it is being served. Therefore, it is not as straightforward as simply stating the preference of condiment. So, in this case, the preferred condiment would be one primary explanatory, and another explanatory would be the food on which it is being served, which influences the preference of the condiment. Similarly, in the case of the growth rates, their values depend very much on the sludge retention factor because a very low sludge retention factor wherein insufficient sludge is retained will not grow sufficient biomass to utilise substrates albeit at high growth rates. This observation proved to be important from the analysis which follows below.

With this analysis, although the β values of the main effect primary explanatories (those primary explanatories which are not paired with any term to account for interaction) may be low, it does not indicate their non-importance because it includes interaction terms. For this scenario, identifying the most important explanatories becomes slightly more complex and lasso (least absolute shrinkage and selection operator) feature selection allows the model to be made easier to interpret by removing explanatories which do not add any information (Fonti, 2017). This was implemented in R (R Core Team, 2019) with the glmnet package (Friedman, Hastie & Tibshirani, 2010).

Lasso is a semi-discrete regularisation technique and shrinks or excludes explanatories based on the penalty factor which can take values between 0 and infinity. Smaller penalty values result in less exclusion or shrinkage. The benefit of lasso above other continuous regularisation approaches is in its exclusion property because it simplifies the model making it easier to infer important explanatories. More information on this can be found in Appendix C and in Hastie, Tibshirani and Friedman (2008)

Apart from the output variables previously identified to have good correlation coefficients with only the inclusion of the parameters, three additional output variables had an improved coefficient of determination $R^2 > 0.7$, by virtue of the inclusion of interaction, which is shown in Table 17 below.

Table 17: Output Variables with Improved Coefficient of Determination when Interactions between Parameters are Included

Variable	Adjusted R^2
S_H	0.74494
S_NH	0.70088
pH	0.72395

The result of a linear model with interaction is 276 (23 choose 2 for each of the primary explanatory terms with interaction plus an additional 23 terms which are not pairwise) terms each with its' own β coefficient compared with just 23 in the case without interaction. This makes the analysis more complex to identify important primary explanatories and therefore these linear models were subjected to the lasso method in order to simplify the linear model, thereby, making it easier to identify the most important explanatories. However, with the lasso method, because the analysis drops certain terms based on their importance, the adjusted R^2 may also decrease, but this is a small decrease for a much smaller, simpler model.

The initial lasso analysis used penalty factors between 0 and 1000 with larger factors applying greater shrinkage and exclusion. The step size employed was in increments of 5. If, after running this lasso analysis, only two or three options exist with a penalty factor (λ) of 0 (so all explanatories are included), a second penalty factor (above 5) and possibly a third larger one with very minimal explanatories or no other penalty factors apart from 0, this means that the increment size is too large, and there may be more possible penalty factors in between 0 and 5. So, the lasso was now repeated for a range of 0 and 10 with increments of 0.05 which allows a greater range of possible models to be identified. Table 18 and Table 19 summarises the effect of the step size on the lasso results. The lasso results include adjusted R^2, Akaike Information Criterion (AIC) and Bayes Information Criterion (BIC). AIC and BIC are model selection criteria that seek to minimise error relative to a "true" model (Chakrabarati & Ghosh, 2011). Consequently, lower values of the criteria are preferred.

Table 18: Lasso results for S_NH with Increments of 5 for Penalty Factors

λ	Number of Explanatories	Adj.R2	AIC	BIC
0	23	0.70088418	109616.0367	111695.5577
5	4	0.689205694	109851.7025	109919.0251
10	3	0.689026829	109857.24	109909.6021

Table 19: Lasso results for S_NH with Increments of 0.05 for Penalty Factors

λ	Number of Explanatories	Adj.R2	AIC	BIC
0	23	0.70088	109616	111695.56
0.05	23	0.69637	109610	110156.1
0.1	21	0.69384	109699	110095.37
0.15	21	0.69337	109716	110090.03
0.2	20	0.6933	109717	110076.17
0.25	20	0.69337	109712	110055.81
0.3	20	0.69336	109713	110065.01
0.35	19	0.69344	109705	110019.26
0.4	18	0.69358	109696	109987.82
0.45	16	0.64555	111598	111852.61
0.5	18	0.6456	111601	111884.79
0.55	16	0.64513	111614	111868.15
0.6	16	0.64507	111616	111870.4
0.75	15	0.64509	111614	111861.14
0.8	14	0.6449	111619	111851.21
0.9	14	0.64493	111617	111841.74
1.05	12	0.64353	111667	111876.35
1.1	11	0.64349	111666	111860.78
1.15	10	0.64354	111662	111841.88
1.2	9	0.64352	111661	111825.7

λ	Number of Explanatories	Adj.R2	AIC	BIC
1.35	9	0.64354	111659	111816.55
1.45	8	0.64254	111694	111836.16
2	8	0.64255	111693	111827.25
2.05	8	0.64257	111692	111826.74
2.4	8	0.64258	111691	111817.82
2.5	7	0.64231	111698	111810.69
2.7	7	0.69053	109804	109931.11
2.75	6	0.69033	109810	109922.32
2.9	6	0.69033	109809	109913.76
3.15	5	0.69016	109815	109912.6
3.45	5	0.69016	109814	109904.06
3.55	5	0.68959	109837	109919.61
3.6	4	0.68921	109852	109919.03
5.6	3	0.68903	109857	109909.6

After running the lasso method, the adjusted R^2 is now 0.68903, but the linear expression has been reduced to be only dependent on i_N_org_mol_perC (nitrogen content of the organisms), mu_AD and f_{ret}, an interactive term with mu_AD and f_{ret} and a second interactive term between f_{ret} and i_N_org_mol_perC. Upon further analysis of the model without interaction, it was evident that f_{ret} and mu_AD were the most significant in this model too with the R^2 coefficient of 0.6349, and therefore inclusion of the interaction term allows the R^2 coefficient to be improved by 0.054104 which is not a large improvement, but at the same time, the model with the interaction as well as the lasso allows for the verification that no other interactions are significant. The observation makes sense because for longer sludge ages (set by higher f_{ret}) more sludge forms which result in fewer ammonia species in the bulk liquid. The same pattern was observed for S_H and pH, and detailed graphs are in Appendix D

5.5.3.4 Linear Models using Parameters and Variables

The next attempt to improve the adjusted R^2 coefficient, for the linear models of the remaining output variables, involved including other secondary explanatories as part of the linear model. This was deemed acceptable because, for many of the output variables, their values are not only influenced by primary explanatories but also by other secondary explanatories, for example, the inhibiting effect of pH on VFA accumulation. For all of the remaining output variables analysed (those which did not obtain a high R^2 in the analyses above with only primary explanatories), all obtained an adjusted R^2 coefficient greater than 0.7. However, the model is complex to analyse because there are so many terms, and thus, lasso was implemented.

In the case of propionate, for example, the full model includes 38 terms in the expression each with its own β coefficient and an adjusted R^2 of 1. The initial lasso analysis (with bounds of 0 and 1000) resulted in penalty factors which were multiples of 5. Therefore, the analysis was repeated using the smaller increment size of 0.05. The simplest model identified with this analysis dropped 33 terms with the adjusted R^2 now being 0.996564. However, the resulting simplest linear model contained only secondary variables which indicate an association between the output variable and the secondary variables but not causality. Although it may point to causality, such as for the example low pH can cause S_VFA to increase; there is no way (at this stage of the investigation with limited statistical scope) of determining if the relationship is one of association or causality.

In order to overcome this obstacle, the linear model, which included primary explanatories as well as secondary explanatories, was selected instead. The model, although not the simplest one, is still considerably simpler than the model with all explanatories included. When considering the parameters only of this information, it can provide information on the most important parameters, provided that the data is standardised. This specific model had a penalty factor (λ) of 0.35. The explanatories are provided in Table 20 below. The results of all the sensitivity analyses are provided in Appendix D. Within the appendix, the graphs which indicate the β coefficients and adjusted R^2 for the revised models obtained with the lasso method (if applicable) are shown.

Table 20: Explanatories, Penalty factors and Adjusted R^2 for Output Variables

Output Variable	Explanatories	Lasso penalty	Adj R^2
Alkalinity	KS_AM, K_I_H2, Y_AD b_AC, b_AM, b_HM, i_N_Org_mol_perC, i_O_Org_mol_perC, mu_AC, mu_HM, S_CO3, S_Glu, S_H, S_H2, S_NH, S_Pr, X_AC, TSS, p_H	0.1	0.999
S_CO3	Y_AD, b_AC, b_AD, i_H_Org_mol_perC, i_N_Org_mol_perC, mu_AD, Alkalinity, S_Glu, S_H, S_NH, S_Pr, X_AC, X_AM, X_HM, Gasflow_CO2	0.1	1.00
S_H2	K_I_H2, b_AC, f_{ret}, mu_AC, mu_HM, S_H, S_Pr, X_AC, Q_GasflowG_CH4	0.1	0.858413
S_Pr	KS_AD, b_AC, f_{ret}, i_N_Org_mol_perC, mu_AD, mu_HM, Alkalinity, S_CO3, S_H, S_H2, S_NH, X_AC, X_HM, Q_GasflowG_CH4, TSS, p_H	0.45	1
S_VFA	KS_AC, KS_AD, KS_AM, KS_HM, K_I_H2, K_I_H_AD, Y_AC, Y_AD, Y_AH, Y_AM, Y_HM, b_AC, b_AD, b_AM, b_HM, f_{ret}, i_H_Org_mol_perC, i_N_Org_mol_perC, i_O_Org_mol_perC, mu_AC, mu_AD, mu_AM, mu_HM, Alkalinity, S_CO3, S_Glu, S_H, S_H2, S_NH, S_Pr, X_AC, X_AD, X_AM, X_HM, Gasflow_CH4, Gasflow_CO2, TSS, p_H	0.00	0.926163
X_AC	KS_AC, KS_AD, KS_HM, K_I_H2, K_I_H_AD, Y_AC, Y_AD, Y_AH, Y_AM, Y_HM, b_AC, b_AM, b_HM, f_{ret}, i_H_Org_mol_perC, i_N_Org_mol_perC, i_O_Org_mol_perC, mu_AC, mu_AM, mu_HM, S_Glu, S_H, S_H2, S_NH, S_Pr, S_VFA, X_AD, X_AM, X_HM, TSS	5.00	0.700
X_AD	Y_AD, b_AD, f_{ret}, Alkalinity, S_Glu, Gasflow_CO2, TSS	9.65	0.997901323
X_AM	Y_AD, Y_AM, b_AM, f_{ret}, S_CO3, S_Pr, TSS	5.3	0.99468
X_HM	Y_HM, f_{ret}, Alkalinity, S_NH, TSS	5.00	0.989449
TSS	Y_AD ,Y_HM, b_AD, b_AM, i_O_Org_mol_perC, mu_HM, S_H2, S_NH, S_Pr, X_AC, X_AD, X_AM, X_HM	1.9	0.99967

A difficulty arises because it is unclear whether it is appropriate to make such an inference regarding the important parameters because the secondary explanatories association is more apparent than the primary explanatory causality. Further, in lasso, when attempting to minimise the error in the lasso objective function by testing different penalty factors, it does not necessarily "remember" the previous tested penalty factors explanatories. This is one of the criticisms of lasso over other discrete methods such as forward or backward selection wherein it may not explore every single possible model. However, the reason for the selection of lasso over these other methods was the large number of possible models to be explored in the discrete methods which add a significant time component in order to obtain the results.

The alternative would be to have a linear model with **only primary explanatories without meeting the required adjusted R^2 coefficient of 0.7** in which case not a large amount of the variability can be explained by the statistical model. However, this is analysed with lasso again in order to attain if the primary explanatories identified in the full model still feature in the analyses with no secondary explanatories. So, in the above example, propionate (S_Pr) was considered. Thus, the results of a linear model wherein only primary explanatories are included can be compared against a model wherein the variance in the data can be explained significantly, but with the issue that secondary explanatories allow for most of the variance to be accounted for, but do not indicate causality. The results of a linear model for propionate (S_Pr) with only primary explanatories is shown in Table 21 which can be compared and contrasted with that in Table 20. It is evident that the same important primary explanatories were identified despite a low adjusted R^2 coefficient. The remaining linear models including their beta coefficients for the remaining output variables with only primary explanatories are given in Appendix D (Subsection 1). In certain instances, it was found that the model which includes secondary explanatories does not include f_{ret} as an important primary explanatory while the linear model which does not include secondary explanatories includes f_{ret} as possibly the most important primary explanatory. The reason for this stems from the effect of other secondary explanatories having such a large association with the output variable in question, so much so that it overshadows the possible causality of f_{ret} in the linear models. This further motivated the need for a comparison between the linear model which includes and also excludes secondary explanatories. Further, this also highlights the consequences of simply implementing the results from statistical analyses without an understanding of the system.

Table 21: Lasso results for S_Pr with only primary explanatories

Output Variable	Explanatories	Lasso penalty	Adj R^2
S_Pr	f_{ret}, mu_AD, mu_HM	7.3	0.213789687

As a final task of establishing the important parameters, a global sensitivity analysis with the CSTR WSR-UASB model confirmed that the effluent acetate and propionate concentrations were most sensitive to the half-saturation concentrations (K) and not the maximum specific growth rate (μ). Using the WSR-UASB model for this sensitivity analysis was permissible because the sludge distribution in the bed does not significantly affect the effluent acetate and propionate concentrations and the WSR-UASB is not subject to the sensitivity of the initial masses like the WI6-UASB.

In summary, the statistical analysis found that the maximum specific growth rates, i.e. mu_AD, mu_AC, mu_HM and mu_AM, were the most important primary explanatories and were subject to the value of f_{ret}. Though f_{ret} may not always have been identified as important for each output variable, the correlation plot given in Figure 14 highlighted f_{ret} as the primary explanatory which is the most correlated one indicating its' importance. The global sensitivity analysis identified the KS_AC and KS_AM as the important primary explanatories required to match the effluent concentrations as measured by Sam-Soon et al. (1990). Although the yield values were identified as important primary explanatories for the biomass concentrations, the biomass concentrations were not measured, and hence, the yield values could not be calibrated at this stage.

Lastly, as an indication of the complexity of the problem at hand, consider a system wherein glucose is fed, but this forms methane gas very quickly such that the intermediates like propionate, acetate and hydrogen do not accumulate. In this scenario, the linear correlation could easily be fit just by inclusion of primary explanatories (parameters) only and not considering any interactive terms or secondary explanatories because glucose, carbon dioxide and methane gas production were modelled with more than 70% of the variability explained in the initial model fitting containing only primary explanatories. However, with the inclusion of intermediates, the variability in the data cannot be accounted for by only considering the primary explanatories, which then brings on the complexity of causality and association. In this complex case, this is where the benefit of steady-state models lie. In the simplification of the problem by assuming steady-state conditions, the equations thus derived

95

allow greater understanding of the relationships at hand without delving into the statistical aspects, in which many researchers in the field do not have expertise.

5.5.3.5 WI6-UASB Analysis

Linear models for the output variables of the WI6-UASB system did not yield any reliable results because of the problem with generating a reliable dataset. This is a model which has the 23 parameters for the case above plus additional retention factors for ADR2 to ADR5. The variables are the same as above but for five reactors resulting in 80 variables. This gives a total number of terms (for each linear model without interactions) of 106 (1 variable is excluded from this count since the effect of the variable on itself is never considered). A more detailed analysis which includes interactions, as well as secondary explanatories, requires a large dataset in significant excess of 5565 runs because this is the number of possible 2-way interactions between the terms as well as the explanatories which are not pairwise. Therefore, a large dataset was attempted to be generated. However, due to the manner in which WEST® runs the analysis, one is not at liberty to choose the primary explanatory values, nor the initial masses after the analysis has begun. The instability of the WI6-UASB is due to the initial masses - if they are too far from the final steady-state value which is impacted mostly by the retention factor, the run stops before steady state is reached. Since one cannot edit the initial masses for each sensitivity analysis run, the system fails to run to steady state, and this was the case in most runs. In a 100-run simulation, between 5 and 10% of the runs resulted in reliable values while the remainder of the runs had failed (resulting in a value of 0). The computational time for this is extremely long. So regrettably, this aspect could not be further investigated. However, because the above sensitivity analysis completed for the WSR-UASB, was not based on target values, the observed linear models explain the phenomena in the W6S-UASB reasonably well too.

5.5.4 Morris Screening Analysis

The Morris Screening method (Morris, 1991) is a one-at-a-time approach wherein one parameter is varied at a time while the remaining ones are held constant. The effect of the variables based on these changes are evaluated, and the mean and standard deviation of the elementary effects are calculated. The elementary effects are computed by considering the value of a variable with the change in one parameter and deducting from that the value of the variable with no changes in any parameter then dividing by the change applied to the parameter. The mean of the **absolute** elementary effects is often used instead to avoid the problem with the effect of opposite signs in computing the elementary effects (Campolongo, Cariboni & Saltelli, 2007). If the mean is less than an established threshold value, it can be considered to be non-influential and so do not require calibration. The line $2\sigma\sqrt{r}$ where r is typically between 10 and 20 as recommended by Campolongo, Cariboni and Saltelli (2007) is the boundary which determines the effect of explanatories on the variable. If an explanatory lies below this line, the effect is linear while explanatories above this line have a non-linear effect on the output or they are involved in interactions. In other words, high σ values point to non-linearity or interactions. The results of the Morris Screening are presented in graphs in Appendix D.

For S_VFA (see Figure 15), Y_AM and Y_AH have high mean values as well as high σ values which indicates a presence of non-linearity and/or interaction. This observation was not found in the SRC method, which only identified the yield values as important for the biomasses such as X_AD, X_AC, X_AM and X_HM for which insufficient data is available to calibrate. One of the most prominent observations with the Morris Screening was that f_{ret} was identified as an important parameter with high mean and mu_AD had a high mean and high σ for many variables which indicates its' possibility of interaction, which was also shown in the linear regression models.

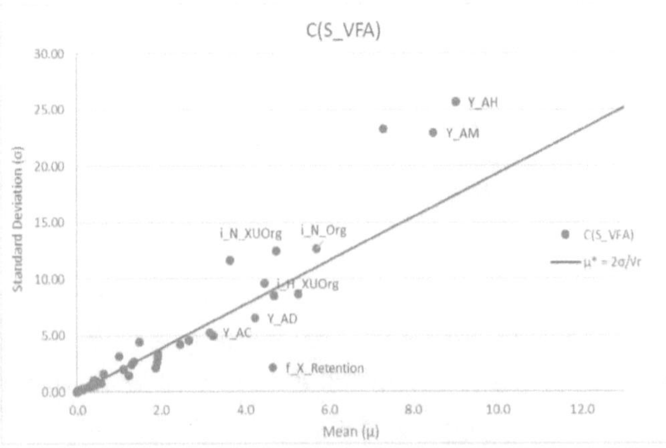

Figure 15: Morris Screening Results for S_VFA

The Morris method, while it is more resilient to Type I errors (the erroneous identification of an explanatory being influential when it isn't), can be prone to Type II errors (failing to identify an explanatory as important) (Ruano et al., 2011). Furthermore, the identification of important explanatories is qualitative instead of quantitative, i.e. the effect which the explanatory has on the output variable cannot be assessed. So, this method was used more as a comparison for the analysis completed using SRC and lasso rather than as a standalone analysis. The Morris Screening method confirmed that most parameters are non-influential except for the maximum specific growth rates and f_{ret}.

5.6 CALIBRATING THE UASB MODEL

5.6.1 Effect of the Default Kinetic Parameters

5.6.1.1 Single Reactor UASB Models

Prior to calibrating the kinetic parameters (Monod maximum specific growth rate, μ and half-saturation concentration, K) of the four AD organism groups following hydrolysis, it was required that the effect of the default values of these parameters be assessed. This was done in order to gauge the magnitude of the parameters and their effect on the effluent concentrations of glucose, acetate and propionate, which provides insight into the manner in which they can be improved. These default values were selected by Sötemann et al. (2005b) to prevent intermediate product accumulation in single completely mixed sewage sludge anaerobic digester and they were tested against the influent of Sam-Soon et al. (1990) to verify this.

The single reactor UASB system was set up in PWM_SA_AD in WEST® (WSR-UASB, Figure 12) with the default values (see Table 23). The initial biomass masses obtained with the SSSM were then entered into WSR-UASB as starting values for the simulation. The WSR-UASB model predictions of the effluent glucose concentration corresponded very closely with those calculated by SSSM and measured by Sam-Soon et al. (1990). However, the calculated effluent concentrations of acetate and propionate were much lower than the values measured by Sam-Soon et al. (1990), due to the high maximum specific growth rates and low half saturation coefficients. Nevertheless, the results predicted by the WSR-UASB system indicated that:

> (1) The SSSM generated sufficiently good starting biomass masses to serve as input for the WSR-UASB and

(2) The VODE mathematical solver was sufficiently stable to generate relatively rapid solutions. This moved the project to a point where the W6S-UASB and WI6-UASB models could be run.

The results predicted by the WSR-UASB system is given in Table 22 below.

Table 22: Effluent Concentrations of the WSR-UASB with Default PWM_SA_AD Kinetics

Component	Concentration (g/m^3)
S_Glu	443.54
S_VFA	0.122
S_Pr	10.995

5.6.1.2 Multiple Reactors UASB Models

In order to run the WI6-UASB system, an approximate starting distribution of the total sludge mass between the five reactors of the series needed to be known (the 6th reactor represented the supernatant above the sludge bed with no sludge in it). In the SS5M, the default values for the maximum specific growth rates (μ) and half-saturation concentrations (K) were selected. Then the SRT of each reactor in the series was changed such that the predicted glucose concentration exiting the reactor matched the concentrations measured by Sam-Soon et al. (1990) at that height up the UASB. The steady-state biomass masses so obtained were then entered into WI6-UASB as starting sludge distribution for the simulation, and the SRT for each reactor from SS5M was converted to a retention factor (f_{ret}).

Figure 11 shows the WI6-UASB configuration. The outflow from ADR5 flows into a secondary settler which is coded to operate as a point settler. This means that solids are either classified as settleable or non-settleable and those that are non-settleable exit with the overflow while settleable material exits the settler with the underflow. All solids were assumed to be settleable. The waste flow is abstracted from the underflow of the point settler while the clear overflow enters ADR6. In this way, the top of the sludge bed was maintained at ADR5 and the supernatant above the bed at ADR6 in the WI6-UASB modelling set-up. As a starting point for simulating the integrated WI6-UASB system, again the default values for the maximum specific growth rates (μ) and half-saturation concentrations (K) for each ADR in the series were selected in order to observe the trend of the acetate and propionate concentrations up the UASB. The μ and K values are global parameters, and so the same μ and K has to be specified for the stacked ADR1 to ADR5.

In running the WI6-UASB model with the default growth kinetic parameters (μ and K), the Runge-Kutta 4 with Adaptive Stepsize Control (RK4ASC) (described by (Press, Flannery, B.P. Saul A. Teukolsky & Vetterling, 2007)) numerical solver had to be used to prevent simulation crashes, despite good estimates for the initial masses. This solver was significantly more stable than other solvers such as VODE (Brown, Byrne & Hindmarsh, 1989) for the six-reactor in-series integrated UASB system. Furthermore, the VODE solver can at times yield inaccurate results due to the step size which it employs. This makes it challenging to observe a concentration trend when using the VODE solver. However, the RK4ASC took much longer to complete the simulation. So, to reduce runtimes, the separated W6S-UASB model, in which the outflow of an upstream ADR is manually set as the influent to a downstream ADR, was run with the RK4ASC numerical solver. The separated W6S-UASB model run with the RK4ASC solver yielded much faster total runtimes for the six reactors of the series than the integrated WI6-UASB model run with RK4ASC. Because both the separated W6S-UASB and the integrated WI6-UASB yielded the same results, the separated W6S-UASB model was run with the RK4ASC solver in further calibration work.

As a starting point for running the separated W6S-UASB system, again the default values for the maximum specific growth rates (μ) and half-saturation concentrations (K) were selected for all the reactors in the series in order to observe the trend of the acetate and propionate concentrations up the UASB. Using the SS5M model to determine the starting sludge distribution for the W6S-UASB and the default kinetic parameters, the glucose, acetate and propionate concentrations profiles in the UASB were calculated with W6S-UASB. Although it was possible to match the glucose concentration by changing the retention factor (f_{ret}), other intermediate products (acetate, propionate, hydrogen) did not accumulate, and a high hydrogen partial pressure was not predicted (Figure 16 and Figure 17). This indicated that not only did the specific growth rates (μ) of the acetogens and methanogens need calibration, but also the half-saturation concentrations (K) because these controls the product effluent concentrations when the product concentrations are low (at the top of the bed, ADR5). Table 23 shows the concentrations of the products in the six reactors of the UASB, and the retention factors (f_{ret}) for the default maximum specific growth rates (μ) and half-saturation concentrations (K) simulated with the W6S-UASB modelling approach.

The measurements conducted by Sam-Soon et al. (1990) were soluble COD, acetate and propionate. A consistent trend in the datasets of Sam-Soon et al. (1990) was a large concentration of residual soluble COD after deducting the residual acetate and propionate. While initially it was assumed that the residual soluble COD is residual glucose, glucose is well-known to be readily biodegradable and unlikely to remain unutilised. The best assumption, therefore, was that since the seed biomass was grown on glucose and apple juice substrate (Sam-Soon et al., 1987), the organisms are very specific as to the substrates they are able to utilise, and if there are some additional soluble organic by-products formed, they will not get utilised in this UASB. Because it was not clear what the reason was for the high residual COD material, it could not be included in the model. In the model, it can either be considered to be residual microbially generated inert organics or the residual COD in the experimental measurements was unutilised glucose due to some inhibitory substance(s).

In Figure 16, where it was assumed that all the measured soluble COD material was glucose, the predicted and measured glucose concentrations match reasonably well in the lower part of the UASB but it was not possible to model the high glucose concentration measured from about 300mm up the UASB. The acetate and propionate concentration are poorly matched due to the high growth rates of acetoclastic methanogens and acetogens, respectively, as shown in Figure 17. These required further calibration evident by the results plotted.

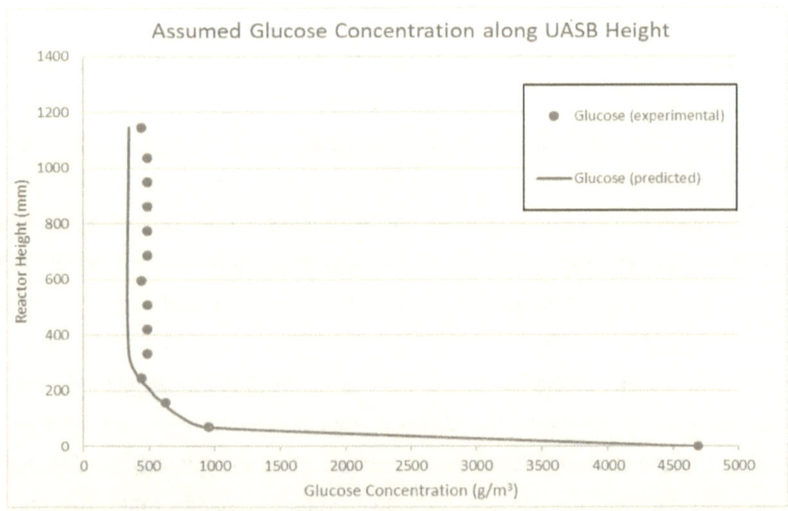

Figure 16: Glucose Concentration along the Height of the UASB for Default Growth Rates.

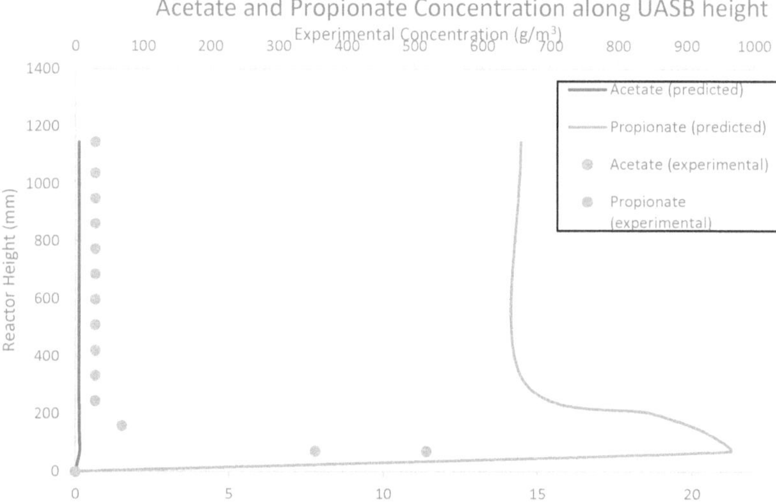

Figure 17: Acetate and Propionate Concentration along the Height of the Reactor for Default Growth Rates.

Table 23: Predicted Concentrations of Glucose, Acetate, Propionate and Hydrogen (Figure 16 and Figure 17) and the Determined Retention Factors with the Default Specific Growth Rates (μ) and Half-saturation Concentrations (Ks).

Reactor	Max specific growth rate (mu) (/day)	Ks (g/m³)	Glucose (g/m³)	Acetate (g/m³)	Propionate (g/m³)	Hydrogen (g/m³)	f_{ret}
ADR6			352.19	0.12	14.45	0.00	0
ADR5	AD = 0.8	AD = 0.781	352.19	0.12	14.45	0.30	0.2
ADR4	AC = 1.15	AC = 0.089	513.36	0.13	18.74	0.34	0.4
ADR3	AM = 4.39	AM = 0.013	608.12	0.14	19.98	0.36	0.5
ADR2	HM = 1.2	HM = 0.156	723.14	0.14	20.82	0.37	0.75
ADR1			953.65	0.14	21.22	0.39	0.97595

5.6.2 Calibrating the Identified Important Parameters

The statistical analysis identified the maximum specific growth rates, retention factors and half-saturation coefficients to be the most important parameters in the model. By using the default growth rates in PWM_SA_AD, it was shown that they are too high to match the measurements of Sam-Soon et al. (1990). Because the half-saturation concentrations (K) govern low bulk liquid concentrations, these were changed first to give better predictions for the effluent acetate and propionate concentrations. The sensitivity analysis completed above confirmed that the half-saturation coefficients are the significant parameters at the measured effluent acetate and propionate concentrations. The parameter estimation tool in WEST© was used to determine the K concentrations for the acetogens and acetoclastic methanogens. So, the K concentrations for the acetogens and acetoclastic methanogens that gave the best match to the effluent acetate and propionate concentrations were selected as the most appropriate. Table 24 lists the values for the half-saturation concentrations (K) so determined for the acetogens and acetoclastic methanogens.

Because Sam-Soon et al. (1990) did not measure the hydrogen concentration, the K value for the hydrogenotrophic methanogens could not be determined. Also, in the experimental UASB, the residual soluble COD concentration could not be confirmed to be glucose, and thus it was not evident if glucose had a low bulk liquid concentration or not, so the default half-saturation constant for the acidogens utilising glucose was retained.

Table 24: Half Saturation Constants Used

Organism Group	Symbol	Half Saturation Constant (mol/m^3)
Acidogens	X_AD	0.781 (Sötemann et al., 2005b)
Acetogens	X_AC	2 (determined with WSR-UASB)
Acetoclastic Methanogens	X_AM	2.5 (determined with WSR-UASB)
Hydrogenotrophic Methanogens	X_HM	0.156 (Sötemann et al., 2005b)

Once the half-saturation concentrations (K) were determined, the separated W6S-UASB model was used to determine the maximum specific growth rate values for the four AD organism groups. The system initially needed to be simplified in order to find a domain of stable model behaviour. Therefore, the effect of pH on the methanogens was switched off in order to restrict the rate of accumulation and utilization of acetate on the specific growth rates only without pH effect. Because glucose utilisation is predominantly influenced by the retention factor and the maximum specific growth rate (mu_AD) of the acidogens (the minuscule amount of glucose formed in the hydrolysis of biodegradable organics formed by the death-regeneration process had a negligible effect), the acidogen µ was selected from literature values and served as the basis for the remainder of the calibration project. The mu_AD value (for the acidogens) was set to the lowest reported value found in the literature (0.4/day) and the corresponding retention factor was determined in order to match to the assumed glucose concentration at Port 1 (corresponding to ADR1) as measured as soluble COD less acetate and propionate by Sam-Soon et al. (1990). This selected specific growth rate was also applied to the remaining digesters in the series.

In the subsequent compartments of the UASB series, it was found that the selected maximum specific growth rate for the acidogens was far too high to predict the measured glucose concentrations. Because the WEST® model was predicting much lower glucose concentrations than measured (if it is assumed that residual COD less acetate and propionate is glucose), the lowest possible growth rate for the acidogens resulted in the lowest error. This justified selecting 0.4 /day for the maximum specific growth rate of the acidogens.

In the experimental UASB, the acetate and propionate concentrations begin to decrease from a height of 70mm, which is the outflow of ADR1 up the UASB series. The reduction in propionate indicates that the acetogens have begun to utilise propionate, which, from the AD model, means that the hydrogen partial pressure has decreased to a value low enough for the acetogens to become active again (they are predominantly inactive under high hydrogen partial pressure conditions). On this basis, the maximum specific growth rate of the hydrogenotrophic methanogens (mu_HM) was used to match the predicted and measured propionate concentration in ADR1, and the maximum specific growth rate of the acetogens (mu_AC) was used to match the predicted and measured propionate utilisation in ADR2 to ADR5. This required in an iterative procedure between ADR1 and ADR2 where (1) mu_HM was set to match the propionate concentration in ADR1. The effluent of ADR1 was fed to ADR2, and in ADR2, mu_HM was set at the same value as ADR1. The mu_AC value was then changed to match the propionate concentration ADR2. Once a value for mu_AC was determined for ADR2, the mu_AC in ADR1 was set equal to it, and a new mu_HM for ADR1 was determined. This procedure was continued until the mu_HM, and mu_AC values converged with good predictions for the propionate concentration in ADR1 and ADR2.

The acetate accumulation in the bottom of the UASB is governed by the glucose utilisation rate (governed by the maximum specific growth rate of the acidogens - mu_AD), acetate utilisation rate (governed by the maximum specific growth rate of the acetoclastic methanogens - mu_AM) and propionate utilisation rate (governed by the maximum specific growth rate of the acetogens - mu_AC) via the effect of the changing partial pressure of hydrogen between high and low values. Because mu_AC and mu_AD were already determined, only mu_AM needed to be calibrated. Once a mu_AM was found by matching the acetate concentration in ADR1, the same mu_AM was applied to ADR2, and this gave a good prediction for acetate in both ADR1 and ADR2. The three maximum specific growth rates (mu_HM, mu_AC, mu_AM) determined for ADR1 and ADR2 were then applied to all remaining digesters (ADR3 to ADR5), and the retention factors were varied in order to find the best fit for the measured data. Once the retention factors, maximum specific growth rates and half-saturation concentrations were determined with the separated W6S-UASB model, the results were checked with the integrated WI6-UASB model. The same results were predicted.

The predicted and measured glucose, acetate and propionate concentrations up the UASB are shown in Figure 18. The Monod constants and retention factors (f_{ret}) are listed in Table 25 and Table 26, respectively. Although the model does not predict the assumed glucose concentrations accurately, the predicted values for acetate and propionate compare well with the experimental values. Because total soluble COD was measured, and methane was not measured, it is not possible to know what the actual residual glucose concentration is. For the same maximum specific growth rate of the acidogens, to predict the glucose concentration requires altering the retention factors and this results in poor predictions of the acetate and propionate concentrations. It was concluded that it was not possible to predict with the AD model a high residual glucose concentration. This was deemed acceptable because no information was available regarding the percentage of residual glucose of total residual soluble COD.

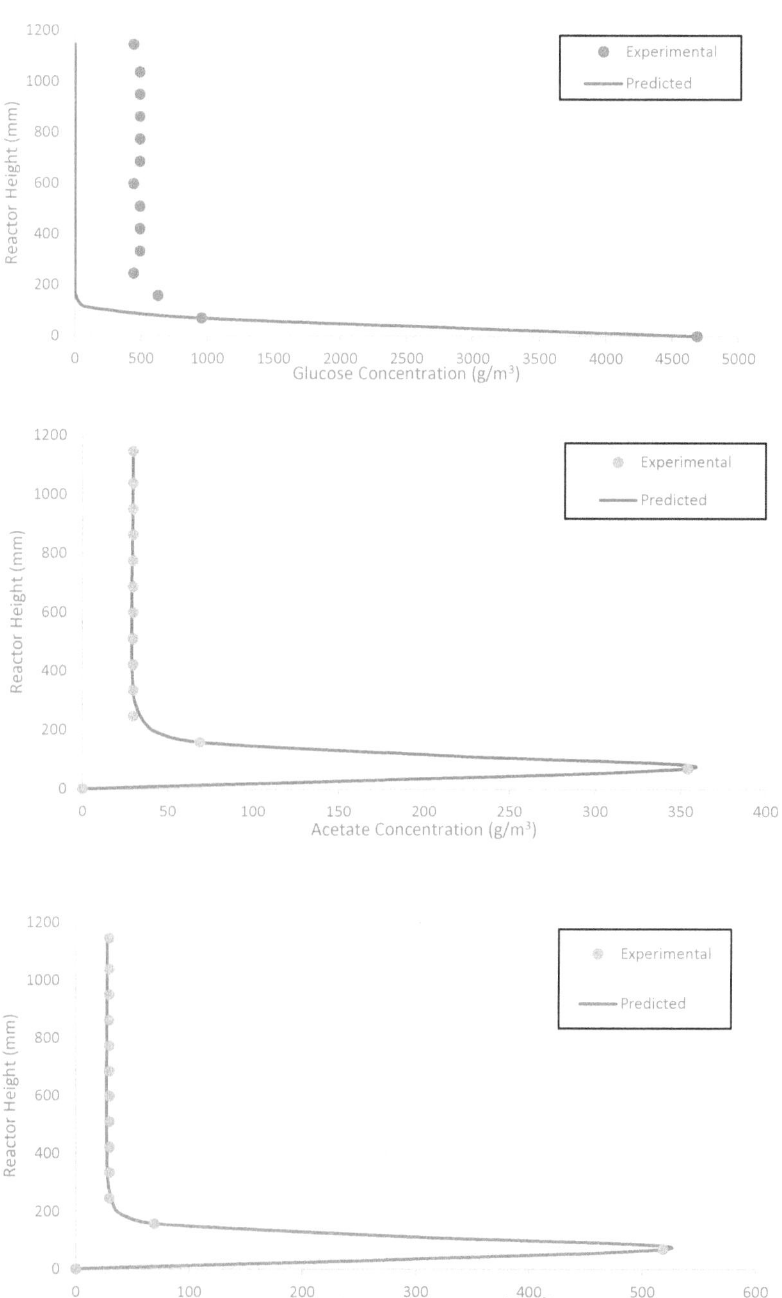

Figure 18: Graphs showing Predicted Concentration of Glucose, Acetate and Propionate along Reactor Height

Table 25: Maximum Specific Growth Rates (μ), Half-saturation Coefficients (K) of the Four AD Organism Groups that predicted the Glucose, Acetate and Propionate Concentrations in Fig 18.

Organism Group	Growth Rate (mu) (/day)	Half Sat Coefficient (Ks) (g/m^3)
Acidogens	0.4	0.781
Acetogens	0.9175	2
Acetoclastic Methanogens	0.49	2.5
Hydrogenotrophic Methanogens	0.3828	0.156

Table 26: Retention factors (f_{ret}) and TSS concentrations of the six in-series UASB system that predicted the Glucose, Acetate and Propionate Concentrations in Fig 18.

Reactor Number	f_{ret}	Waste Flow (m^3/day)	TSS Concentration (g/m^3)
ADR1	0.988725	-	28046.36
ADR2	0.984	-	26135.53
ADR3	0.9785	-	20501.32
ADR4	0.905	-	4663.77
ADR5	0.6	0.0001	475.21
ADR6	0	-	0

5.7 MODELLING THE pH UP THE UASB

All the above simulations were done without pH inhibition of the AD biomass groups. When the predicted and measured acetate and propionate concentrations matched well, it was expected that the pH prediction would also correspond well with the experimentally measured values. However, this was not the case - the predicted pH was far lower compared with that measured by Sam-Soon et al. (1990) (Figure 19). Because aqueous phase speciation usually does not require calibration because the endpoint of the dissociation reactions are thermodynamically defined and reach equilibrium very fast, the too low pH has two possible causes- either there was (i) an error in the aqueous speciation or CO_2 gas expulsion part of the model or (ii) too much acid in the aqueous phase. The latter was excluded because as much as possible CO_2 was expelled – the CO_2 gas expulsion was modelled with aqueous-gas equilibrium reactions based on Henry's law. So, the CO_2 gas expulsion/dissolution equilibrium reactions were carefully checked, and indeed an error was found. When this was corrected, the predicted pH maintained the same trend up the UASB but was now higher than measured by Sam-Soon et al. (1990).

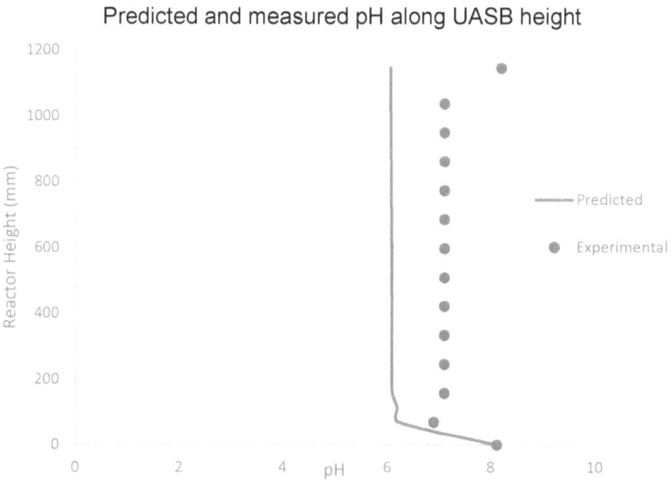

Figure 19: pH Prediction Prior to Debugging and Changes in CO_2 Gas Expulsion Reactions.

Because the lower zones of the UASB are not subject to the headspace gas content, the CO_2 gas expulsion probably is not at equilibrium with the headspace CO_2 partial pressure. Less CO_2 expulsion would reduce the pH. In addition, any gas formed in the lower layers of the UASB would not escape with the gas flow but instead would be entrained in the liquid phase. This aspect of gas entrainment was added into the model. At the top of the UASB, all gas would then exit in the gas phase. Ideally, the gas should escape at the top of the UASB (i.e. from ADR6), but due to the inclusion of a point settler to retain solids within the sludge bed, the gas escapes out of the point settler instead.

Evaporation is a significant process in AD, and water vapour in the headspace has the effect of reducing the CO_2 partial pressure. Therefore, the kinetics of water evaporation was also added to the model. The exiting gas was assumed to be saturated with water vapour, and its pressure was assumed to be approximately equal to the vapour pressure of pure water at the temperature of the anaerobic digester. The Antoine Equation can be used to calculate the vapour pressure, and this is given by Equation (34)

$$P_{H_2O} = 133.33 \times 10^{A - \frac{B}{C+T}} \tag{34}$$

where

P_{H2O} is the water vapour pressure in Pa

133.33 is the conversion factor from mmHg to Pa

T is the temperature in °C

A=8.07131, B=1730.63 and C=233.426

The CO_2 expulsion was changed to a rate-controlled reaction with the expulsion rate specified as model input, and it evolves according to the difference between the equilibrium partial pressure and the headspace partial pressure. The quantity of CO_2 expelled to the gas phase could be set by the kinetic rate - if low, a high concentration of CO_2 remains dissolved which decreases the pH. The equations below describe the CO_2 evolution process which was added into PWM_SA_AD.

$$CO_2 Rate = K_{ev,CO2} \times 1000 \times \left(molality[H_2CO_3] - \frac{P_{CO2}}{101325 \times Henry_{CO2}} \right) \tag{35}$$

$$K_{ev,CO2} = \frac{KLA_{CO2} \times T_{corr}}{MW_{G_{CO2}}} \tag{36}$$

where
$CO_2 Rate$ is the rate at which CO_2 evolves in mol/m³/day
$K_{ev,CO2}$ is the rate constant in /day
1000 converts units from L to m³
KLA_{CO2} is the mass transfer coefficient in g/mol/day
T_{corr} is the temperature correction coefficient
MW_G_CO2 is the molar mass of CO_2 in g/mol
molality[H_2CO_3] is the molar concentration of H_2CO_3 in mol/kg, which is close to mol/L since water is the solvent.
P_CO2 is the partial pressure of CO2 in Pa
101325 converts from Pa to atm

Henry$_{CO2}$ is Henry's Coefficient for CO_2 in L.atm/mol and is adjusted to account for temperatures different to 25°C based on the van't Hoff Equation (Loewenthal, Ekama & Marais, 1989).

The "HeadSpace Pressure" term also was changed to allow it to be different from atmospheric pressure if required. In a tall column UASB, the pressure on the gas bubble is different compared with the pressure at the surface. This is due to the hydraulic pressure caused by the head of water. For the sub-reactors, the hydraulic pressure was taken to be the average of the pressure at the bottom of that reactor and at the top of that reactor. This allows the actual pressure on the bubble to be more accurately represented.

Following these changes and with a high CO_2 expulsion rate to induce conditions close to CO_2 equilibrium as before, the predicted pH followed the same pattern as before with CO_2 gas expulsion equilibrium reactions, i.e. higher than the measured pH observed by Sam-Soon et al. (1990). So, the CO_2 mass transfer coefficient (K_{LA_CO2}) was decreased until the predicted pH matched the measured pH well. The predicted and measured pH profiles in the UASB are shown in Fig 8.

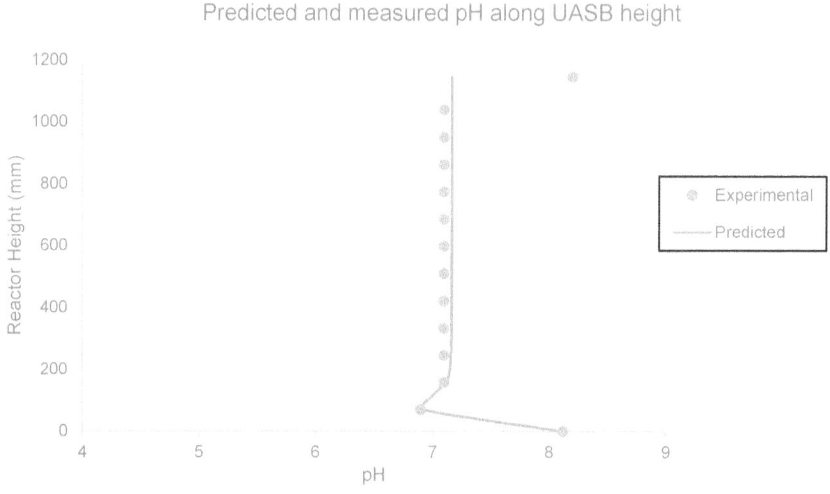

Figure 20: pH Prediction after Changes to Gas Evolution Kinetics and with better calibrated KLA_CO2 Parameter

5.8   MODELLING THE ACETOCLASTIC METHANOGEN pH INHIBITION

At this point, pH inhibition of the acetoclastic methanogens was switched on. The inhibition term used in methanogenesis allows the acetoclastic methanogens to be completely inhibited below a pH of 6, partially inhibited between a pH of 6 and 7 and no inhibition above a pH of 7. The inhibition term included in the kinetics of the growth of the acetoclastic methanogens is given by Equation (37) where pH_{UL} is the pH above which no inhibition occurs (in this case 7), and pH_{LL} is the pH below which 100% inhibition occurs (in this case 6). The inhibition function was obtained from Batstone et al. (2002) since the previous inhibition term used by Sötemann et al. (2005b) did not allow the acetoclastic methanogens to be sufficiently inhibited by pH.

$$I_{pH} = \begin{cases} 1 & pH \geq 7 \\ e^{-3\left(\frac{pH-pH_{UL}}{pH_{UL}-pH_{LL}}\right)^2} & pH < 7 \end{cases} \tag{37}$$

This resulted in an increase in acetate concentration in the bottom of the UASB because the predicted (and measured) pH in this region is below 7 which decreases the maximum specific growth rate of the acetoclastic methanogens. The maximum specific growth rate of the acetoclastic methanogens, mu_AM, was thus increased slightly to restore accurate prediction of the acetate concentration in the bottom of the UASB in order to compensate for now including the pH inhibition effect on the methanogens. The revised growth rate for the acetoclastic methanogens is 0.495/day. All other kinetic constants were not changed.

Finally, once the separated W6S-UASB model was working well with the calibrated kinetic constants and inhibition factors, the integrated WI6-UASB system was run under the same conditions. The integrated WI6-UASB system gave identical predictions - the only difference was the slow simulation runtime. The runtimes of a 100-day simulation were: The RKS4AC simulator for both WI6-UASB and W6S-UASB approximately 146 and 70 minutes, respectively. The VODE solver with W6S-UASB takes 20 seconds for each reactor in the series but cannot be used for WI6-UASB. However, time is taken to export values from one digester to use as inputs for the next digester with W6S-UASB.

5.9   ESTABLISHING THE BEST FIT KINETIC CONSTANTS

It is recognised that the sludge distribution up the UASB reactor depends on the initial μ value selected for the acidogens and that the μ values of the other three AD organism groups are dependent on the resulting sludge distribution (retention factors). Had a different μ for acidogens been selected initially, a different sludge distribution up the UASB would have been obtained and the μ values of the other three AD organism groups also would have been different. So, the calibration constants obtained in this calibration is one set of a number of possible sets to model the UASB. Had the sludge distribution in the UASB been measured by Sam-soon et al. (1990), this degree of freedom would not exist. Importantly, in UASBs, generally the solids are at a higher concentration at the bottom of the bed and at lower concentrations further up. The model predictions matched well with the typical phenomenon of UASB reactors. The predicted solids distribution of the different organism groups is shown in Figure 21.

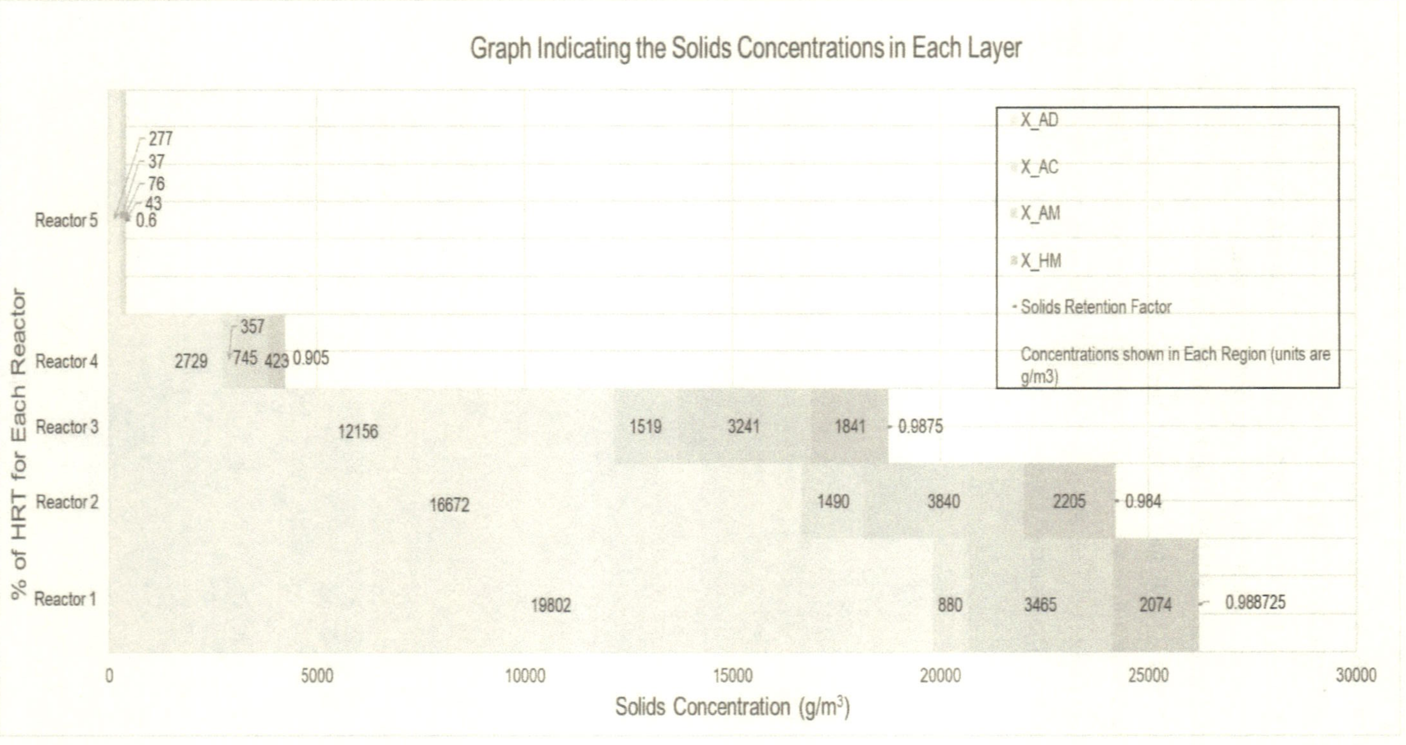

Figure 21: Predicted Biomass Solids Concentration in Each Sublayer

Extra information runs were required to draw information regarding how unique the calibrated parameters are. As a first step, the sludge distribution in the bed was considered. This is a function of the retention factors. From Figure 21 above, the total mass of sludge, including the amount contained in inert material and BPO formed during the death regeneration phase is approximately 34gTSS. Therefore, the specific calibrated growth rates were all doubled, and the initial sludge masses were halved while the retention factors were kept constant to evaluate the effect on the sludge bed.

Because of the doubled specific growth rates, the system appears to be more efficient, and this resulted in lower effluent substrate concentrations. As a result of the lower effluent concentrations, more sludge mass was formed - the total mass of sludge in the digesters increased to 39gTSS. This increase can be attributed to the greater utilisation of substrate and thus if the effluent concentrations had been the same as measured while retaining the same solids retention factors, the total mass of sludge would remain constant at 34gTSS. Therefore, the total sludge mass in the system is, in fact, independent of the specific growth rates selected and rather dependent on the retention factors of the system. The sensitivity analysis supported this wherein it was shown that f_{ret} displayed the highest beta value combined with a low p-value indicating its importance and that it is statistically significant.

So, for fixed retention factors, total TSS remains constant, and for different retention factors, the TSS mass in the sludge bed would be different. Because Sam-Soon et al. (1990) did not measure the sludge bed VSS and TSS, the predicted bed solids concentration could not be validated. Therefore, it was investigated if there was another set of constants that could predict bed behaviour equally well.

For this aspect, the methodology used to calibrate the parameters to the UASB system was repeated for different starting values of mu_AD. This changes the glucose concentrations, which requires an adjustment of the retention factor in ADR1 to predict the concentration accurately. Consequently, for a higher retention factor (or lower), the kinetic specific growth rates of acetogens, acetoclastic methanogens and hydrogenotrophic methanogens require a change as well in order to still predict the bed concentrations of acetate and propionate. The sludge bed mass then also changes which is due to the change in the retention factors.

The calculations were repeated for a range of mu_AD values and consequently different retention factors. Table 27 below summarises the results of the growth rates, while Table 28 provides the corresponding retention factors. It was found for the case where **mu_AD is 1.5/day**, the effluent concentrations out of ADR5 are **fully insensitive to the retention factor**. This is due to the very low concentration entering ADR5 such that all the substrates appear to be utilised very easily. This means that the predictions of Group 5 are poorer than the rest which is indicated by the Coefficient of Determination. Table 29 below reports the Coefficient of Determination for each group of acetate, propionate and pH. Since the gas flow was not measured, glucose concentration could not be determined and thus was not used in evaluating the Coefficient of Determination.

Table 27: Summary of Additional Groups for Growth Rates of Organism Groups following Hydrolysis

Organism Group	Group 1	Group 2	Group 3	Group 4	Group 5
Acidogens	0.40	0.50	0.60	0.80	1.50
Acetogens	0.9175	1.17	1.41	1.93	3.50
Acetoclastic Methanogens	0.495	0.6125	0.746	0.99	1.87
Hydrogenotrophic Methanogens	0.3828	0.4875	0.593	0.80	1.56

Table 28: Retention Factors for the New Groups of Calibrated Constants

Reactor	Group 1	Group 2	Group 3	Group 4	Group 5
ADR1	0.988725	0.985525	0.98235	0.97595	0.9536
ADR2	0.984	0.98	0.9745	0.9655	0.935
ADR3	0.9785	0.97	0.967	0.954	0.92
ADR4	0.905	0.885	0.85	0.7	0.6
ADR5	0.6	0.4	0.3	0.2	0.1

Table 29: Coefficients of Determination for the Groups of Constants

Group Number	Coefficient of Determination (R2)		
	Acetate	Propionate	pH
1	0.9996	0.9999	0.9821
2	0.9995	0.9986	0.9822
3	0.9994	0.9996	0.9821
4	0.9991	0.9995	0.9820
5	0.9975	0.9989	0.9820

Based on Table 29, Group 1 had the largest overall Coefficient of Determination and was selected as the best fit for the remainder of the study. The values obtained for the growth rates of the organism groups, within Group 1, are within reasonable range compared to the literature values, whereas Group 4 and Group 5 constants are significantly higher than reported literature values.

5.10 FITTING OF CONSTANTS TO REMAINING EXPERIMENTS

The above groups of constants were then used in order to predict a different experiment of Sam-Soon et al. (1990). Because minimal information was provided on how the following experiments were conducted, the best possible assumption is that they were conducted under identical conditions as the first experiment. Ideally, the calibrated rates should have been able to predict the system results. However, the difficulties which arose can be attributed to:

- No gas (methane, carbon dioxide and hydrogen) flow measurement, which makes it difficult to establish the amount of actual glucose utilised.

- No sludge wastage and sludge bed measurements details were provided, which makes the identification of a sludge age difficult.

- No initial inoculum seed details provided.

Although the issues presented were the same as that of the first experiment above, the difference is that in that case, the freedom to change the growth rates was still available. However, if it is desired that the growth rates be held constant, the difficulties become much more apparent. Nevertheless, the calibrated constants provided above were used to attempt to predict the results of Experiment 2. The influent information details of this experiment are provided in Table 30.

Table 30: Influent Parameters for Extra Experimental Systems

Parameter	Symbol	Quantity	Unit
Influent Flow	Qi	30	L/day
Total COD (Glucose only)	Sti	2792	mgCOD/L
Influent pH	pH	8.12	-
Influent Alkalinity	Alk	3021	$mgCaCO_3$/L
Reactor Volume	V	9	L
Organic Loading Rate	OLR	9.307	$kgCOD/m^3$/day
Free and Saline Ammonia	FSA	65.8	mgN/L

The Trace Element Solution, as given by Table 12, was the same as for the extra experimental system. Therefore, the only adjustments required to the Influent WEST® file were due to the flow rate, total COD, Alkalinity, pH and FSA. This required the use of the pre-processor Excel file once again in order to speciate into the required components. Once this was completed, the influent file could be generated and fed to the UASB with the same volumes for the sublayers as previously used.

5.10.1 Summary of Experiment 2

The graph shown in Figure 23 indicates the concentrations of acetate, propionate and soluble COD as measured by Sam-Soon et al. (1990) in Experiment 2. Visually, it is clear that the acetate and propionate concentrations are virtually identical in magnitude. What this means mathematically, in order to be accurately predicted, is that the specific growth rates of acetoclastic methanogens and rate of acetogens need to be of similar order of magnitude in combination with a high partial pressure (consequently a low hydrogenotrophic methanogen specific growth rate) thereby directing all the glucose via the path of propionate production with equal acetate production. Alternatively, a lower propionate utilisation rate and higher acetoclastic methanogenesis rate in combination with less glucose being directed via the high partial pressure pathway resulting in less propionate and more acetate

117

being formed from glucose. Because all of the combinations of calibrated parameters in Table 27 above did not follow either of these relationships, it was expected that the calibrated parameters would not result in a good prediction of Experiment 2. This expectation was realised.

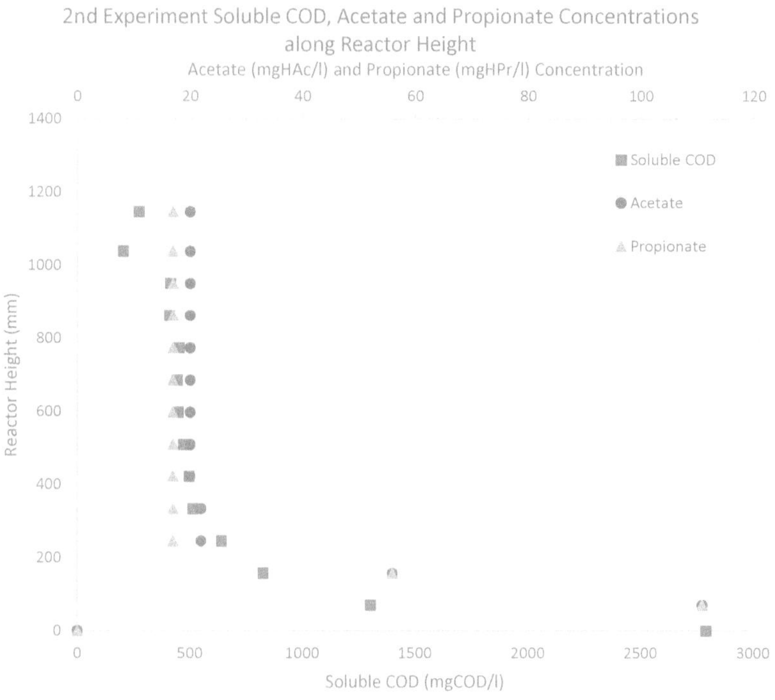

Figure 23: Experimentally Measured Values for Experiment 2

Furthermore, since in general, the growth rates of the propionate utilisers (acetogens) of all the calibrated parameters were much higher than the acetoclastic methanogen growth rates, the predicted concentrations for Experiment 2 had lower propionate concentrations than acetate concentrations, but relative to the glucose concentration, these predicted concentrations were both higher than the measured values. Therefore, the growth rates identified previously could not be applied to Experiment 2.

However, it is possible to identify different specific growth rates which can predict Experiment 2. The value of the model lies in the ability to retain the gas in the lower levels of the UASB, apply a pressure to the gas bubbles which is dependent on the depth of the UASB, retain solids without the use of a settling tank, and allow for gas evolution to not necessarily be at equilibrium with the dissolved concentrations. Without these added benefits and improvements to the model, the temporary failure situation at the bottom of a UASB and subsequent utilisation of the intermediates higher up the UASB would not have been predicted. So, while global growth rates could not be identified, the importance of the model with its modifications is still extremely useful. Table 31 below summarises the specific growth rates for the four organism groups following hydrolysis for the same growth rates of acidogens used previously. The retention factors follow in Table 32.

Table 31: Summary of Growth Rates for Experiment 2

Organism Group	Group 1	Group 2	Group 3	Group 4	Group 5
Acidogens	0.40	0.50	0.60	0.80	1.5
Acetogens	1.2	1.50	1.80	2.50	4.5
Acetoclastic Methanogens	0.88	1.10	1.33	1.76	3.32
Hydrogenotrophic Methanogens	0.425	0.545	0.67	0.90	1.8

Table 32: Retention Factors for the Calibrated constants for Experiment 2

Reactor	Group 1	Group 2	Group 3	Group 4	Group 5
ADR1	0.994589	0.99305	0.9915	0.988425	0.97765
ADR2	0.9895	0.987	0.984	0.9775	0.96
ADR3	0.975	0.968	0.963	0.953	0.9
ADR4	0.96	0.95	0.94	0.92	0.84
ADR5	0.9	0.89	0.88	0.8	0.7

The experimental systems of Sam-Soon et al. (1990) were run with the objective of verifying a hypothesis on the formation of pellets which is very different from the purpose the results are used for this study. Although one global set of constants was not identified, the model has the ability to predict failure once the rates are calibrated to the dataset. An unknown factor was whether or not Experiment 2 was set up in an identical manner to Experiment 1

– the results were not clear. This included questions such as where was the inoculum obtained and was the same quantity of inoculum used in both experiments? Furthermore, the lack of measurements of the gas flow makes it difficult to ascertain the amount of glucose actually utilised and obtain a COD balance over the experimental systems.

5.11 SUMMARY OF CALIBRATION PROCEDURE

From the experience of this calibration work, the protocol for calibrating an AD model to a UASB system with temporary failure at the bottom of the bed is recommended as follows:

- Determine the required volumes for each compartment in the UASB for the six in-series separate (W6S-UASB) and six in-series integrated (WI6-UASB) systems, where the sixth reactor is the supernatant volume above the bed. These volumes are found by determining which points along the reactor heights are significant endpoints (i.e. the point at which acetate and propionate begin decreasing and the point where no further glucose utilisation takes place). If necessary, more reactors in series can be modelled, but this will result in more work and longer run times.

- Use a steady state single reactor spreadsheet model (SSSM) to obtain approximate values for initial masses for a completely mixed UASB. If sludge bed VSS measurements are available, compare the predicted bed sludge mass with that predicted and establish reasons for significant differences.

- Use the initial masses and run a single reactor CSTR simulation in the dynamic model (WSR-UASB) with some default literature maximum specific growth rates and half-saturation constants to examine the system performance with the initial (default) kinetic parameters.

- Use the five in series steady state spreadsheet model (SS5M) to determine the initial masses needed in the W6S-UASB and WI6-UASB systems based on default kinetic parameters. Alternatively, if sludge distribution/VSS profile results are available, then distribute the sludge mass in the sludge bed according to the measured VSS profile.

- Rerun the dynamic AD model with default kinetic values in WI6-UASB or, to avoid slow runtimes, use W6S-UASB. This will indicate whether the specific growth rates and half-saturation constants require calibration or only one of these two groups.

- If the half-saturation constants require calibration, use the WSR-UASB system with the parameter estimation (PE) tool in the dynamic model software to determine the half-saturation constants which will generate model predicted effluent values similar to the measured values.

- Run the W6S-UASB system to determine the specific growth rates. This will require an iterative procedure to converge onto the correct maximum specific growth rate values for the hydrogenotrophic methanogens (mu_HM) and acetogens (mu_AC)

- Test the system with the WI6-UASB and WSR-UASB system models to confirm that the same results are obtained.

- In all simulation runs, ensure that the initial biomass masses in the reactor and sub-reactors are reasonably close to the final predicted masses. Initial biomass masses can be estimated with the spreadsheet models (SSSM and SS5M) if an SRT is estimated.

5.12 CONCLUSION

The UASB model in WEST® required six-in-series, completely-mixed, anaerobic digesters in order to model the sublayers within the experimental UASB accurately. After having set up the influent file for the Sam-Soon et al. (1990) dataset, which involved speciating the influent components, the dataset was fed into the UASB configuration in WEST®. Where the initial conditions within the digesters were too different from the steady-state masses, it caused instability in the six-in-series UASB system within WEST®, which posed a challenge. This warranted the use of steady-state spreadsheet models, set up with derived steady-state equations, in order to identify the required steady-state masses to be used as initial conditions.

With the UASB WEST® model running in a stable manner, a large number of model predictions were generated. These model predictions were used in a sensitivity analysis in linear regression models within R (R Core Team, 2019), which gave insight into the most important model parameters. The sensitivity of the intermediate products proved to be more complex because the variance in the data cannot be explained with the use of parameters only and depend very much on the other variables. However, even with the use of the lasso method to identify the most important explanatories, the association between the variables still overshadows the causality of the parameters. This highlighted the difficulty of too much reliance on statistical sensitivity analysis and indicated the importance of steady-state models with significantly simpler equations allowing the important parameters to be identified with greater ease.

The important parameters were then subjected to a series of steps in order to calibrate them to the dataset. While the glucose concentration could not be calibrated due to insufficient measurement, the acetate and propionate predicted concentrations matched the measured values. Following the calibration, the pH prediction was expected to match the measured values. This was, however not the case, and this mismatch was attributed to the assumption of equilibrium between the gas headspace and bulk liquid, which is inaccurate especially in the lower levels of the UASB. When the gas evolution dynamics were changed to be rate-controlled, the pH prediction improved considerably over the height of the bed.

A challenge arose due to the lack of data for the biomass concentrations in the bed (resulting in an unknown sludge distribution) and the gas flow measurements (giving insight into the residual soluble COD concentration). A degree of freedom now existed within the calibration because of the lack of these measurements. This degree of freedom is governed by the choice of the maximum specific growth rate of the acidogens and consequently, the solids retention factor of the first ADR. However, once these two are set, the maximum specific growth rates of the other organism groups are no longer arbitrary. With the additional measurements mentioned, this degree of freedom of the maximum specific growth rate of the acidogens and the solids retention factor of the first ADR ceases to exist.

The calibration procedure developed and used here allowed for the UASB to be calibrated to a second set of measurement as well. So, while a single set of global constants are not attainable, the calibration protocol allows for ease of calibrating to a UASB dataset which started off as significantly more complex. Following this, the calibrated model was used in Chapter 6 for model applications of failure and digester start-up.

6 MODEL APPLICATION

6.1 INTRODUCTION

AD is an extremely useful means of biological wastewater treatment due to the methane generated in the process which can then be used as an energy source. However, despite the benefits of AD and that it is one of the oldest means of treating wastewater, it notoriously has been known to be susceptible to failure. This is often due to the introduction of toxins to the system or too high organic load. Mathematical models such as PWM_SA_AD allow insight into anaerobic digester performance and also can be used for optimisation purposes. Early warning systems are essential to prevent impending anaerobic digester failure but depend on the model's accurate calibration if it is to be used with greater confidence. The AD model developed and calibrated in this study can be utilised in different scenarios to investigate anaerobic digester failure. An important aspect of anaerobic digesters is their start-up which follows a conservative procedure to avoid failure by overloading too soon. The calibrated model can generate insight for digester start-up which is discussed in this section after exploring the AD failure modes.

6.2 USING THE CALIBRATED MODEL TO PREDICT FAILURE

6.2.1 Effect of Incrementally Decreasing Influent Alkalinity

To examine anaerobic digester failure, the model which was calibrated and discussed in Chapter 5 was used to predict a sustained UASB failure condition. Because Group 1 constants had the largest overall Coefficient of Determination, they were selected for further implementation in these investigations. Under sustained failure conditions, the UASB does not recover higher up the reactor, and VFAs (specifically acetate due to the low pH) accumulate. This failure condition was achieved in the model by incrementally reducing the influent alkalinity until there was insufficient alkalinity to buffer the pH. The pH decreased to a point where the acetoclastic methanogens were inhibited and unable to utilise the acetate sufficiently to increase the pH in the upper layers of the UASB. Figure 24 shows the acetate concentrations in the UASB for decreasing influent alkalinity from 6000 to 4200 mg/L as $CaCO_3$. It can be seen that at an influent alkalinity of 4200mg/L as $CaCO_3$, the acetate concentration increases and does not decrease again further up the UASB. At an influent alkalinity of 4200 mg/L as CaCO₃, Figure 25 shows that the pH remains below 7.0 at ~6.48 and does not recover back to >7.0 along the UASB height.

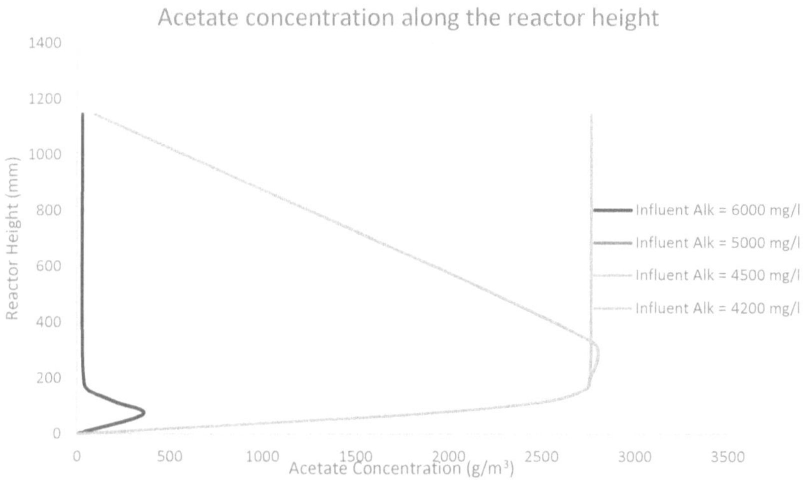

Figure 24: Graph Indicating Accumulation of Acetate as Reactor Approaches Failure

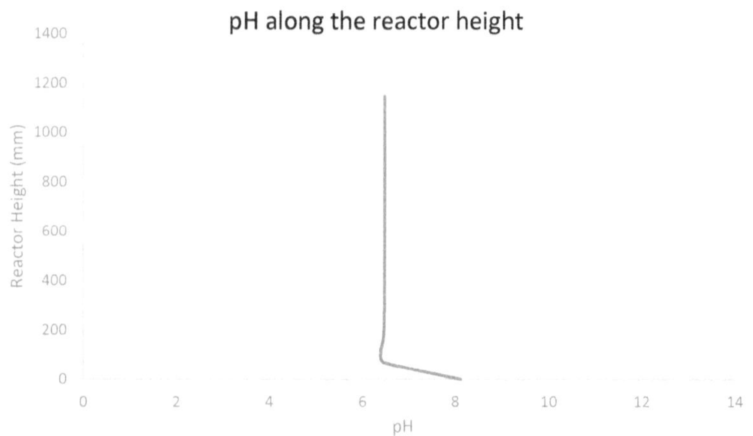

Figure 25: Graph Indicating the pH for the Digester Failure Scenario

6.2.2 Effect of Acetogen Inhibition on Failure

A second scenario of anaerobic digester failure was investigated. In the AD model, high hydrogen concentrations inhibit the acetogenic reaction by means of a non-competitive inhibition function. With this function, the acetogens are never fully inhibited but rather are predominantly inactive at high hydrogen concentrations with the result that propionate is not converted to acetate. With the default inhibition constant ($K\_I\_H_2$) at 10 gCOD/m^3 in the model (obtained from (Sötemann et al., 2005b)), the acetogens are 100% active at a hydrogen concentration of 0 g/m^3 but their activity reduces quite rapidly such that they are less than 10% active at a concentration of 12 gH$_2$/m^3. The acetogens can only utilize the propionate when the hydrogen partial pressure (modelled using concentrations and not actual H$_2$ partial pressure) again decreases. The effect of the ($K\_I\_H_2$) inhibition constant on the acetogens and anaerobic digester failure was investigated by gradually decreasing it. This has the effect of making the acetogens more sensitive to the hydrogen concentration while at the same time, in acidogenesis, the high partial pressure reaction is favoured over the low partial pressure reaction resulting in more propionate (together with acetate and hydrogen) being formed from the substrate acidified by the acidogens.

It was expected that the propionate concentration would increase up the bed and not be utilised at all due to the increased sensitivity of the acetogens to hydrogen. This high propionate concentration was expected to decrease the pH to such a point where the acetoclastic methanogens would be inhibited. However, while it was seen that more propionate accumulates in the bottom of the bed as $K\_I\_H_2$ is decreased, the system was able to recover higher up the bed as the propionate is utilised significantly in the 6th digester. Figure 26 and Figure 27 shows the pH and propionate concentrations respectively up the UASB for $K\_I\_H_2$ decreasing from 10 down to 1 gH$_2$/m^3. Even with $K\_I\_H_2$ =1 gH$_2$/m^3, the acetogens were still able to utilise the propionate once the hydrogen concentration was low enough at the top of the bed. Following the decrease in the propionate, acetate also experiences a decrease in concentration.

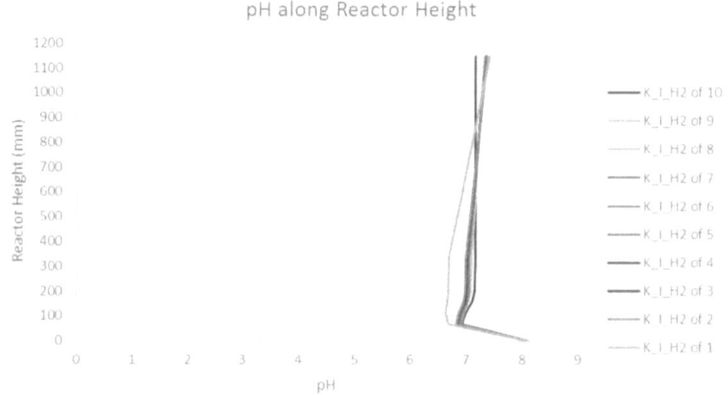

Figure 26: Graph showing pH changes along Reactor Height as K_I_H2 is Decreased

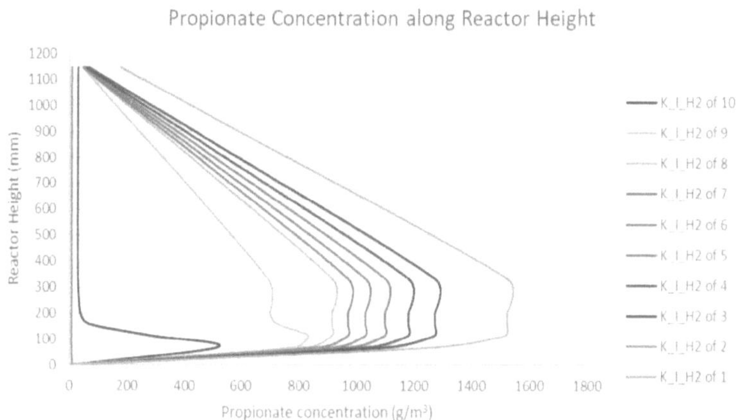

Figure 27: Graph showing Propionate Concentration change along Reactor Height as K_I_H2 is Decreased

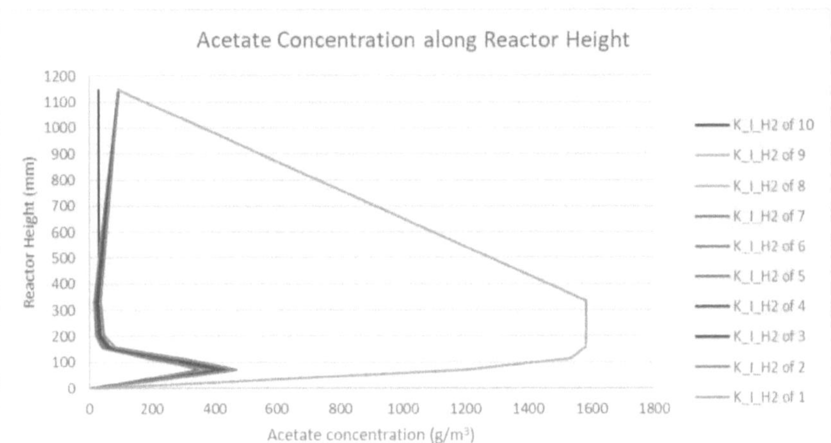

Figure 28: Graph showing Acetate Concentration change along Reactor Height as K_I_H$_2$ is Decreased

The results indicated in Figure 28 above indicate that although the acetate concentration does accumulate along the reactor height, the system is able to recover as soon as the propionate is utilised. Because the inhibition of the acetogens results in its contribution to the acetate concentration not occurring concurrently with the glucose contribution to acetate when K_I_H$_2$ is 1 gH$_2$/m^3, this decreases the acetate load on the acetoclastic methanogens at a specific time. In effect, it may prevent the prediction of anaerobic digester failure to a certain extent. So, the delayed acetogenesis serves to spread the acetate load on the acetoclastic methanogens more evenly up the UASB and "cushion" the acetoclastic methanogens from pH inhibition, thereby preventing system failure. Whether or not this happens, in reality, could not be established.

Failure due to propionate accumulation will only take place if the hydrogenotrophic methanogens are inhibited and cannot utilise the hydrogen to lower the hydrogen partial pressure. Inhibition of the hydrogenotrophic methanogens could either be due to a toxin in the system or due to pH values lower than 6 (Batstone et al., 2002), who report that (i) hydrogenotrophic methanogens are fully inhibited below a pH of 5, and partially inhibited between a pH of 5 and 6 and (ii) acetoclastic methanogens are fully inhibited below a pH of 6 and partially inhibited below a pH of 6 and 7. Thus, if the pH were to drop below 6, failure will be caused by an accumulation of acetate due to inhibition of the acetoclastic methanogens rather than failure due to an accumulation of propionate in response to high hydrogen partial pressure.

To validate this, the influent alkalinity was gradually lowered while still maintaining a value of 1 for K_I_H$_2$ (the lowest tested value) which results in the acetogens being more sensitive to hydrogen accumulation. It was found that with an influent alkalinity of 4000mg/L as CaCO$_3$, the acetoclastic methanogens cannot utilise the acetate and high concentrations exit the system. Figure 29 and Figure 30 show the acetate and propionate concentrations along the reactor height for varying influent alkalinity values respectively and Figure 31, the pH. The influent alkalinity, which causes failure is 4000mg/L as CaCO$_3$ compared with 4200mg/L as CaCO$_3$ in the case where propionate is less sensitive to the hydrogen concentration.

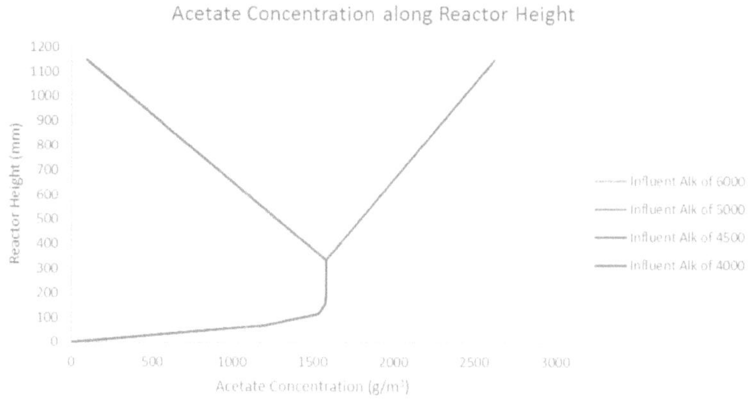

Figure 29: Graphs showing Acetate changes as Alkalinity is Reduced for a Reduced K_I_H2 value

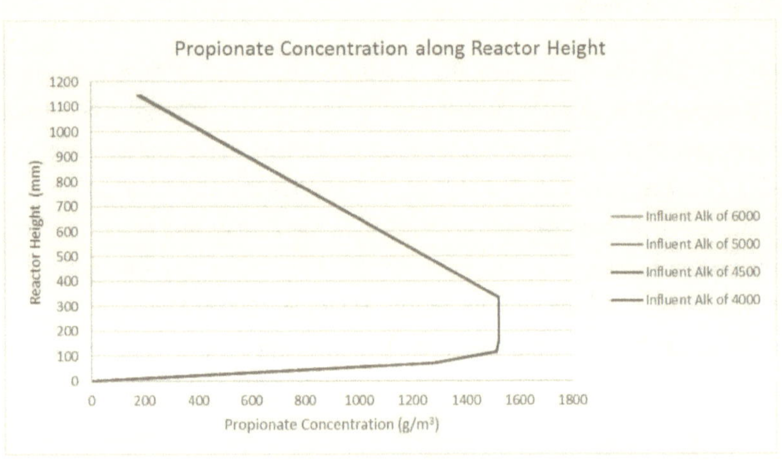

Figure 30: Graphs showing Propionate changes as Alkalinity is Reduced for a Reduced K_I_H2 Value

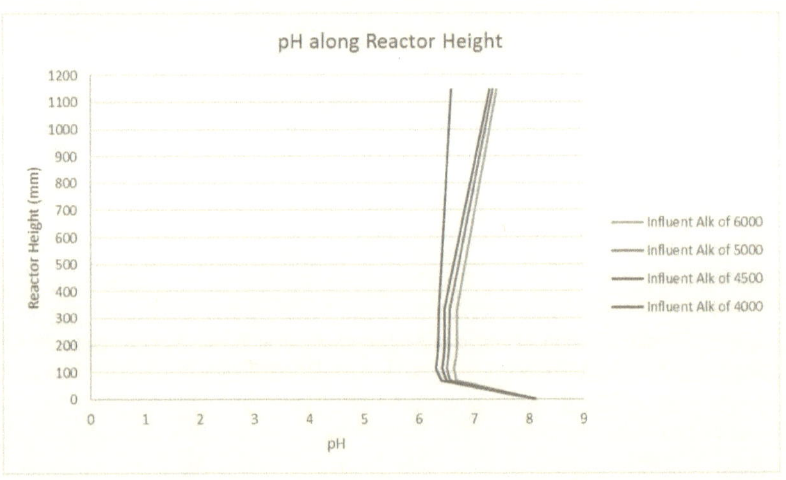

Figure 31: Graphs showing pH changes as Alkalinity is Reduced for a Reduced K_I_H2 Value

From the above two anaerobic digester failure scenarios, it is clear that the PWM_SA_AD model, because it is now better calibrated, has the ability to predict digester failure. The first scenario investigated indicates that the pH prediction is significantly improved with the changes made to the gas evolution kinetics and that the methanogen inhibition is working well for pH values less than 7. The second scenario investigated highlights the importance of the inclusion of hydrogen partial pressure. If the acetogens are made more sensitive to hydrogen concentrations by lowering the inhibition coefficient, the propionate does not get utilised lower in the bed, and thus the acetate load on the acetoclastic methanogens is less than what it would have been if the acetogens were less sensitive to the hydrogen concentration.

6.3 PREDICTING FAILURE WITH ADM1

The biological reactions of ADM1 were coded into PWM_SA_AD as a subset in order to obtain a comparison of how ADM1 predicts failure compared to PWM_SA_AD. All inhibition terms, kinetic rates and stoichiometric equations for ADM1 were obtained from Batstone et al. (2002). The reason for coding in ADM1 into PWM_SA_AD is twofold. Firstly, the format of ADM1 is very different from PWM_SA_AD, and so it is not easy to make a process by process comparison. Secondly, it was important that the pH prediction method was the same in both cases in order to maintain consistency. This required recasting ADM1 into the same components as PWM_SA, which are aligned with the standard equation chemistry convention (Brouckaert, Brouckaert & Ekama, 2019b). Thus, with ADM1 coded into PWM_SA_AD as a subset, the same external speciation routine could be used for both AD models so the performance of the AD bioprocesses could be compared. The way in which ADM1 is coded into PWM_SA_AD is that if ADM1 is in use, all bioprocesses of PWM_SA_AD model are turned off and vice versa.

The same reactor configuration used for the UASB calibration was used for testing ADM1. This meant that all retention factors which were found previously for PWM_SA_AD were also retained for ADM1. The objective of this exercise was to compare how ADM1 performs compared with PWM_SA_AD, and not to calibrate ADM1 to the data of Sam-Soon et al. (1990). The table below indicates the maximum specific substrate utilisation rate (k_m) and the half-saturation coefficients (k_s) for each organism groups. The maximum specific substrate utilisation rate is used in the original ADM1 model, and this was retained for

PWM_SA_AD, but the necessary conversion terms like the yield value were used to make it compatible with PWM_SA_AD.

Table 33: Substrate Utilisation Rates and Half Saturation Coefficients for various organism groups in ADM

Organism Group	Maximum Specific Substrate Utilisation Rate (k_m) (kgCOD.m^{-3}substrate/ kgCOD.m^{-3}organism mass.day^{-1}	Half Saturation Coefficient (k_s) (kgCOD.m^{-3})	Yield Coefficient (Y) (gCOD biomas/gCOD substrate)
Monosaccharides degraders	30	0.5	0.1
Propionate degraders	13	0.1	0.04
Acetate degraders	8	0.15	0.05
Hydrogen degraders	35	0.000007	0.06

Table 34: PWM_SA_AD Maximum Specific Growth rates, Half-saturation Coefficients and Yield Values

Organism Group	Maximum Specific Growth Rate (day^{-1})	Half Saturation Coefficient (k_s) (mol.m^{-3})	Yield Coefficient (Y) (gCOD biomass/g COD substrate)
Acidogens	0.4	0.781	0.0895
Acetogens	0.9175	2	0.0397
Acetoclastic methanogens	0.495	2.5	0.03925
Hydrogenotrophic methanogens	0.3828	0.156	0.04

The results in Figure 32 (glucose, acetate and propionate concentrations), Figure 33 (hydrogen concentration) and Figure 34 (pH) were found for ADM1 with an influent alkalinity of 6000 mg/L as CaCO$_3$.

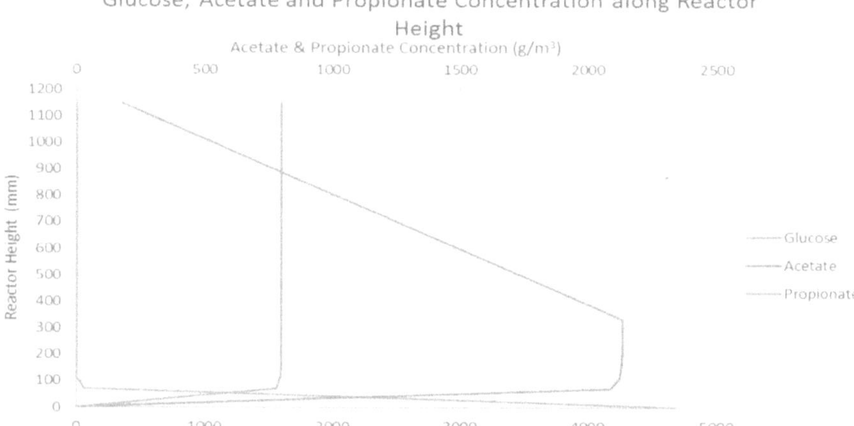

Figure 32: Glucose, Propionate and Acetate Concentration along reactor height as predicted by ADM1

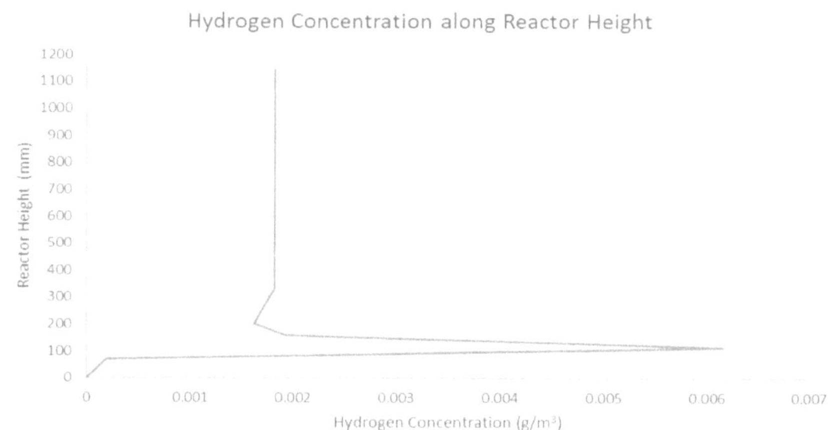

Figure 33: Hydrogen Concentration along Reactor Height as predicted by ADM1

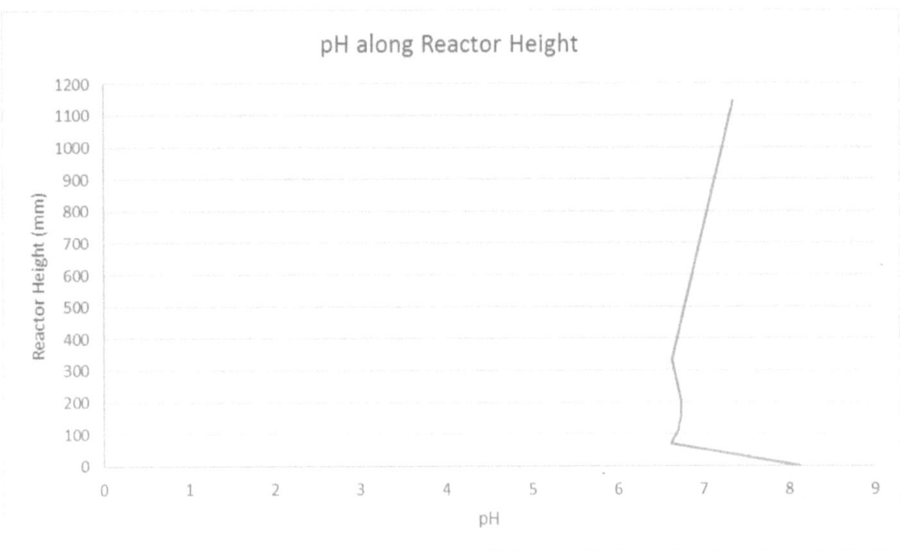

Figure 34: pH along Reactor Height as predicted by ADM1

It can be seen from Table 33 that with a k_m value of 30 for the monosaccharide degraders, and the yield value of 0.1 gives a maximum specific growth rate of 3 /day. This is 7.5 times higher than the value (0.4/day) which was calibrated to fit the data of Sam-Soon et al. (1990) meaning that in this case, the glucose is used far more rapidly than in the calibrated scenario, and in addition, a much larger acetate load is placed on the acetate degraders (acetoclastic methanogens). Similarly, with the hydrogen utilisers, the hydrogen does not accumulate significantly since the equivalent maximum specific growth rate for this organism group is 2.1 (yield of 0.06). This converted k_m value is 5.49 times higher than the value (0.3828/day) that was calibrated to the data of Sam-Soon et al. (1990).

The propionate utilisers are substantially more sensitive to the hydrogen concentration compared with their sensitivity in PWM_SA_AD. ADM1 uses a K value of 0.0035 gCOD/m^3 compared with 10 gCOD/m^3 used in PWM_SA_AD. In addition, the equivalent maximum specific growth rate for the propionate utilisers is 0.52 /day (yield of 0.04) which is 1.76 times **lower** than the value (0.9175) found by calibration to the data of Sam-Soon et al. (1990). The combined effect of the lower maximum specific growth rate of the acetogens in addition to their greater sensitivity to the hydrogen concentration results in an accumulation of propionate, which is not utilised up the reactor bed. However, as shown earlier with PWM_SA_AD, the accumulation of propionate does not significantly impact the pH so that it is still able to recover once the acetate concentration has decreased.

Lastly, for the acetate utilisers, the equivalent maximum specific growth rate in ADM1 is 0.4 /day (yield of 0.05) which is comparable to the value of 0.495/day found with PWM_SA_AD calibration. Therefore, it would be expected that the acetate accumulation would be similar to the pattern observed by Sam-Soon et al. (1990) (and PWM_SA_AD). However, it is not, and the large difference can be attributed to the rapid utilisation of glucose which places a larger initial load on the acetate utilisers compared with that observed by Sam-Soon et al. (1990). Since the pH did not drop too far below 7.0, the acetate utilisers were able to consume acetate at the top of the bed. So, ADM1, while showing a significantly different behaviour up the UASB bed than that observed, does correctly predict non-failure of the UASB system.

Following this, ADM1 was made to simulate failure by reducing the influent alkalinity gradually. As in the case of PWM_SA_AD, it was also found that failure occurs due to an accumulation of acetate which cannot be utilised due to their inhibition by the pH being too low. For ADM1, this occurred at an influent alkalinity of 5000 mg/L as $CaCO_3$ compared with 4200mg/L as $CaCO_3$ found for the calibrated PWM_SA_AD. The reason for this is most likely due to the lower equivalent maximum specific growth rate of the acetate utilisers compared with PWM_SA_AD which results in them being fully inhibited at higher influent alkalinity than was predicted by PWM_SA_AD. At this stage, it would not be possible to say which of the two models is more appropriate to model UASBs and anaerobic digester failure because ADM1 kinetic rates are not calibrated to the same dataset as PWM_SA_AD kinetic rates. This just serves to show that structurally both models can predict failure apparently equally well.

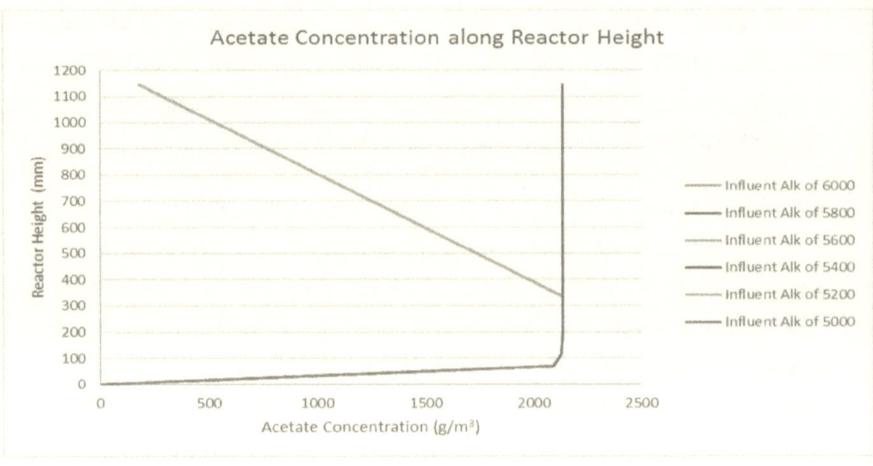

Figure 35: Acetate Concentration as predicted by ADM1 under Reducing Alkalinities

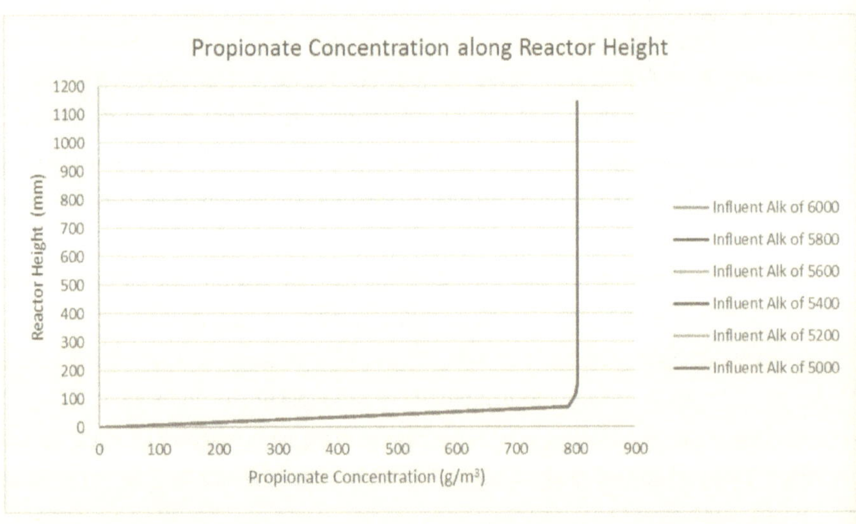

Figure 36: Propionate Concentration as predicted by ADM1 under Reducing Alkalinities

135

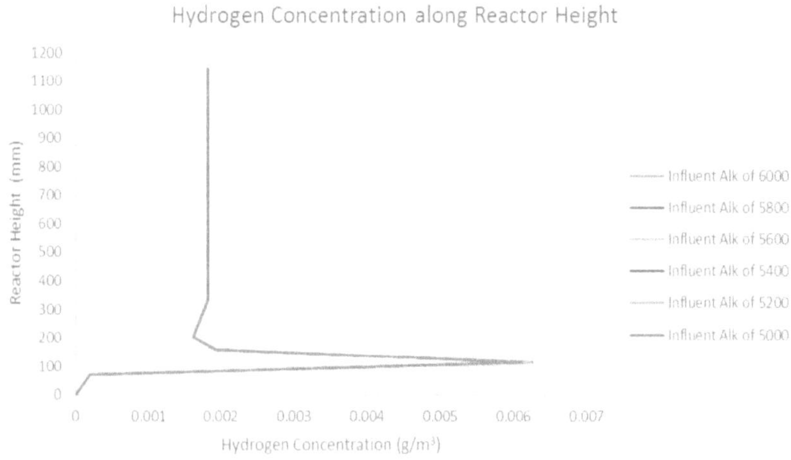

Figure 37: Hydrogen Concentration as predicted by ADM1 under Reducing Alkalinities

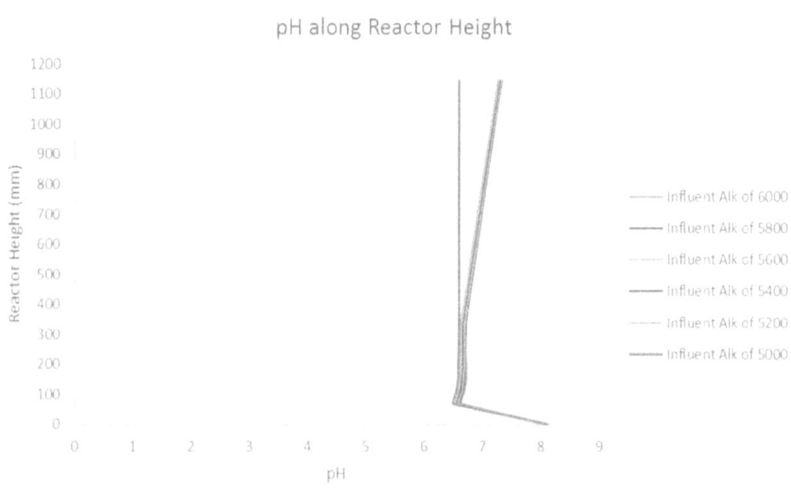

Figure 38: pH as predicted by ADM1 under Reducing Alkalinities

6.4   USING THE CALIBRATED PWM_SA_AD MODEL FOR DIGESTER START-UP

6.4.1 Background

The benefits of AD were discussed in Chapter 2 above, but many complications arise in digester start-up. Frequently, in practice, digester start-up is not well understood, so sufficient alkalinity is fed to the digester upon start-up in the hope that the anaerobic digester will not fail. With a better-calibrated model, it may be possible to simulate a digester start-up using an appropriate controller. For anaerobic digester control, the warning indicators for impending anaerobic digester failure described in Section 2.1.4 are typically used as measured responses (outputs) of the system.

For calibrating a digester start-up scenario, feeding glucose would not be appropriate because, in reality, digesters are fed with primary sludge comprising mostly slowly hydrolysable BPO. A primary sludge dataset was required and that of Izzett (1992) was selected since a detailed analysis and calibration of hydrolysis kinetics had already been completed by Sötemann et al. (2005b).

Sötemann et al. (2005b) discussed different formulations which can be used to model the hydrolysis process, namely first order, first order specific, Monod and surface-mediated (saturation or Contois). Ristow et al. (2005) also provided a detailed comparison. The first-order kinetics described the hydrolysis process reasonably well for both the Izzett (1992) and O'Rourke (1968) datasets in Sötemann et al. (2005b) and Ristow et al. (2005). However, Sötemann et al. (2005b) commented that intuitively the hydrolysis rate formulation should also be dependent on the concentration of the acidogen biomass since these are the organisms which mediate the process, especially for the more complicated kinetic models rather than the simpler steady-state models. Furthermore, Vavilin, Rytov and Lokshina (1996) found that the hydrolysis of cellulose material specifically, which is a particulate material, was poorly predicted with the Monod equation. The main criticism of the Monod kinetics is that it was originally developed for soluble substrates and expresses the substrate concentration with respect to the bulk liquid. This may be acceptable for modelling dynamic conditions of minor disturbances to an established long-term steady-state condition but not for anaerobic digester start-up where both substrate and biomass concentrations with respect to the bulk liquid are very low, but maximum biomass growth rates are expected. In such instances, the saturation kinetics are best because, for this, the growth rate defining condition is the substrate to biomass concentration ratio which would be high even when the bulk liquid concentration is low. Thus, the surface-mediated saturation kinetics were accepted as more appropriate to model hydrolysis of BPO in anaerobic digester start-up.

Because Sötemann et al. (2005b) calibrated the saturation kinetics constants to the Izzett (1992) data for the hydrolysis process, these were selected for the PWM_SA_AD model as well. The values for the surface-mediated kinetics as given in Sötemann et al. (2005b) and Sötemann et al. (2005a) are shown in Table 35 below. The values which are used in PWM_SA_AD are also shown in Table 35. Numerically, although the values with the same units look very different, the composition of the biomass and substrate in Sötemann et al. (2005a) are with respect to carbon being 5 and 3.5 respectively, while in PWM_SA, carbon is always expressed as 1. Hence the PWM_SA_AD k_m and k_s values are the Sötemann et al. (2005a) values multiplied by 5 and divided by 3.5. Figure 39 shows that if these constants were used in a "batch" reactor where BPO are fed and hydrolysed to glucose in the presence of an acidogen biomass seed, both sets of constants yield virtually identical results.

Table 35: Calibrated Hydrolysis Constants for Izzet (1992) data

	Sötemann et al. (2005a) Mol units	Sötemann et al. (2005a) COD units	PWM_SA_AD COD units
K_m (mol/mol/day or gCOD/gCOD/day)	6.797	5.58	7.968
K_s (mol/mol or gCOD/gCOD)	10.829	8.89	12.695

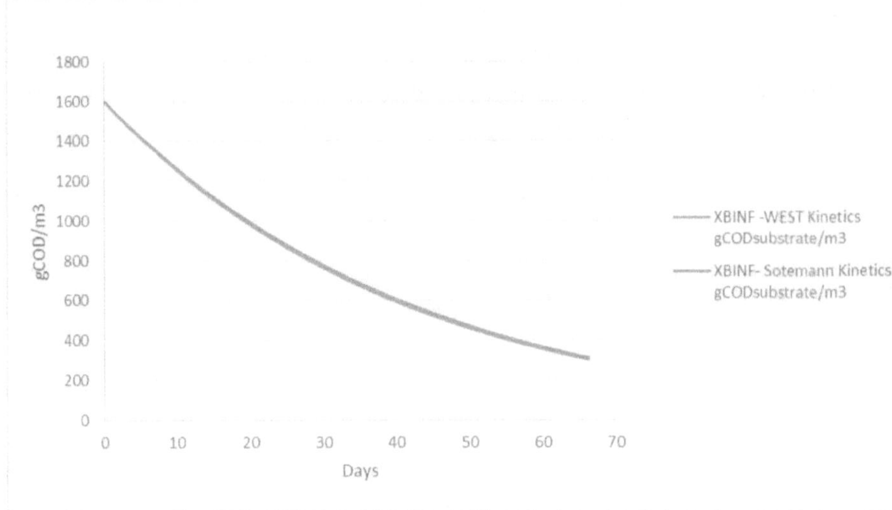

Figure 39: Graph Indicating that Sötemann et al (2005b) Calibrated Constants give identical results t those used in PWM_SA_AD

6.4.2 Prediction of Digester Start-up with Model Prior to Calibration

As a means to evaluate the model prior to calibration, the default kinetic rates (for processes following hydrolysis) were used, and an anaerobic digester was fed with the sludge from the dataset of Izzett (1992). The values which were calibrated by Sötemann et al. (2005b) for the hydrolysis process was used for this process. It was found that even with as little as 1% seed, the digester was able to start up without the need for a controller to regulate the incoming flow. Furthermore, the digester acetate concentration does not accumulate to more than $0.8g/m^3$, while the pH is maintained at greater than 6.5. These observations were made despite the fact that the incoming flow has a pH of only 5.29 and an acetate concentration of 2244 g/m^3 (as HAc). So, it is evident that the model, prior to calibration, does not allow for an accurate representation of digester start-up conditions because the growth rates of the processes following hydrolysis are too high.

The simulations completed hereunder use the calibrated rates to predict digester start-up, and the observations were compared to that of the model predictions prior to calibration.

6.4.3 Overview of Content

The first step required prior to digester start-up is the identification of a suitable influent. The mixture of primary and humus sludge dataset of Izzett (1992) was selected for this purpose. Details of this information follow hereunder in Section 1426.4.4. Once the influent was selected, the start-up procedure was attempted via three separate experimental systems, namely:

1. Start-up with a controller with a constant flow into the system

2. Start-up without a controller but with gradually increasing flow directly into the digester

3. Start-up with a controller with gradually increasing flow into the system.

In cases 1 and 3, where a controller was used, the flow which enters the system is not necessarily the same as the flow which enters the digester at any given point in time. This is because, depending on the measured value of the control parameter within the digester and the setpoint of the controller, the controller may cause some flow to be diverted away from the digester. By contrast, for case 2, the flow into the system will always be the same as the flow into the digester because none of the flow has the option of being diverted. Figure 40 and Figure 41 illustrate the difference between these cases.

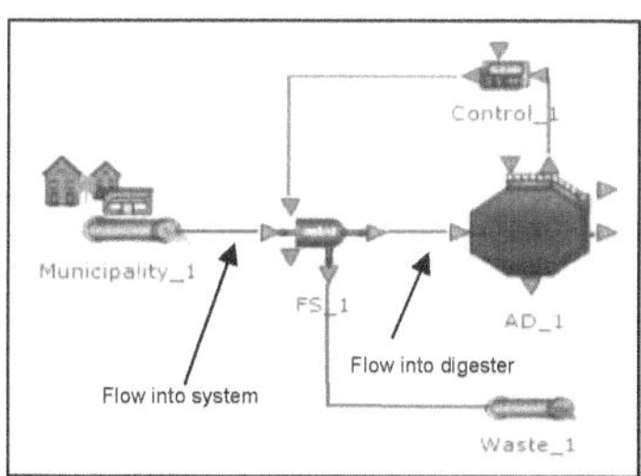

Figure 40: Setup used for Case 1 and 3

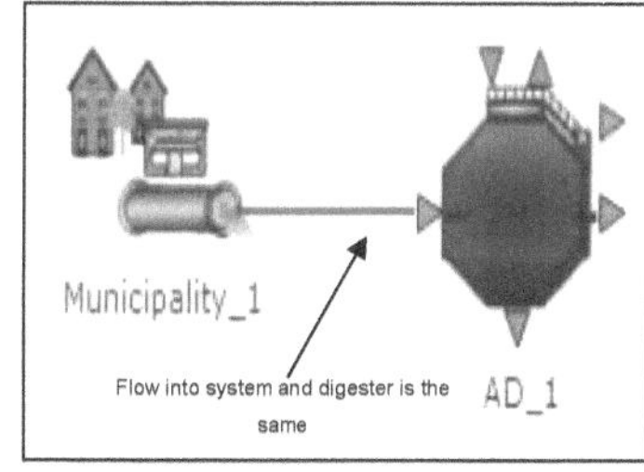

Figure 41: Setup used for Case 2

Investigations into these systems included looking into the selected setpoint value and the factor of proportionality. This is because characteristics between different wastewater treatment plants can differ widely. So, while literature may recommend certain operational limits, these may not be globally applicable.

The hierarchical diagrams which follow give detail as to the information regarding those constants which were changed and those which were held constant at each step in the investigations.

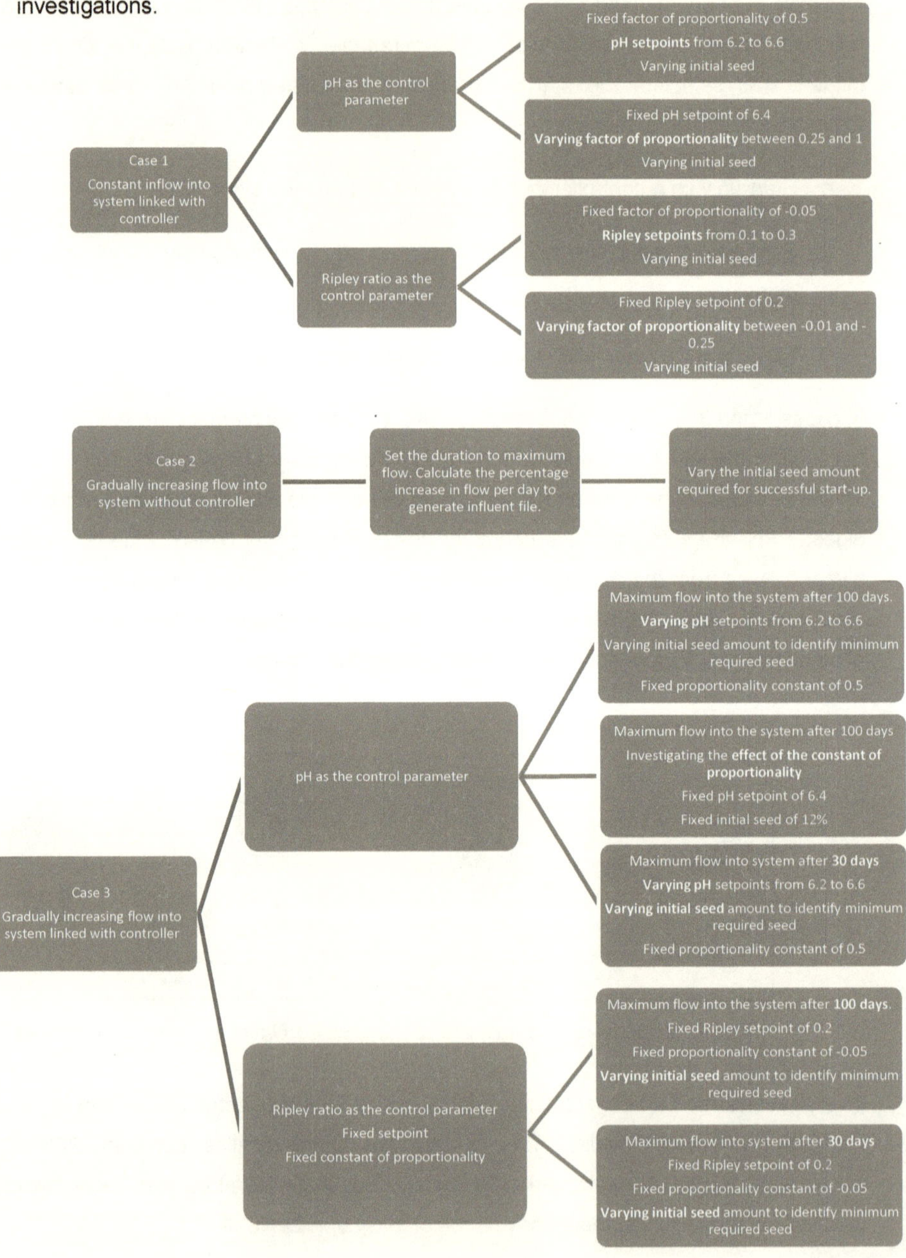

6.4.4 Influent Feeding and Characteristics

Izzett (1992) fed a 14L digester with a mixture of primary and humus sludge and operated it at a sludge age of 20 days with characteristics as shown in Table 36 below. In order to obtain seed with realistic ratios between the different biomass components, this influent was fed to a digester in WEST® with sufficient initial biomass and alkalinity required to run the digester to a long-term steady state. Of this 14L volume, a defined percentage was taken out as the actual seed for a system where a digester is linked to a controller. The remaining volume of this digester was filled with hypothetical wastewater treatment plant effluent wherein the alkalinity was assumed to be 250mg/L as $CaCO_3$ and the pH 7.

This filling of the anaerobic digester with effluent was done here to (i) supply alkalinity and (ii) drive out air from the anaerobic digester to minimize the oxygen content in the anaerobic digester headspace to avoid creating a methane-oxygen gas ratio (5-10%) that can possibly auto-ignite at atmospheric pressure and 600°C (Kundu, Zanganeh & Moghtaderi, 2016).

Table 36: Influent Characteristics of the system operated by Izzett (1992)

Parameter	Value	Unit
Influent flow - Qi	0.7	L/day
pH	5.29	-
Temp	37	°C
Sti	42570	mgCOD/L
TSS	31554	mg TSS/L
Alkalinity	59	mg/L as CaCO3
SCFA	2244	mg HAc/L

6.4.5 Constant Inflow Start-up with a Controller-linked Digester (Case 1)

The central idea of feedback control is that some output response of a system is measured, and this feeds back to a controller which is then used to alter the input into the system. This contrasts with open-loop controllers wherein there is no feedback from the output to the input. Typical controllers which can be used range from very basic ideal (or on/off), proportional (P), proportional plus integral (PI) and lastly, proportional plus integral and derivative control (PID) (Reddi, 2011). The issue with the on/off controller is that the system has no response until the setpoint has been reached. The response, therefore, tends to be aggressive to re-stabilise the system disturbance. The P-controller has the factor of proportionality which allows the system to respond slightly prior to the setpoint being reached. For this case, the factor of proportionality is directly related to the gain. High gain systems switch on and off more frequently and consume more power and consequently, cause greater wear and tear on the equipment (Reddi, 2011). For this reason, the lowest possible gain for the purposes of the desired application should be selected. In reality, a detailed optimisation analysis would be carried out in the controller design. This has not been done here because the purpose is to illustrate the potential of the model to simulate the digester start-up, as quickly as possible without causing anaerobic digester failure.

Figure 42 shows the set-up which was used for the anaerobic digester with a P-controller. This was selected in preference of the On/Off controller because a proportional controller achieves better control and more steady operation. Because a proportional controller is in use, the flow into the digester can be either completely on, completely off or somewhere in between depending on the value of the proportional constant and the setpoint. When the measured value is near or approaching the

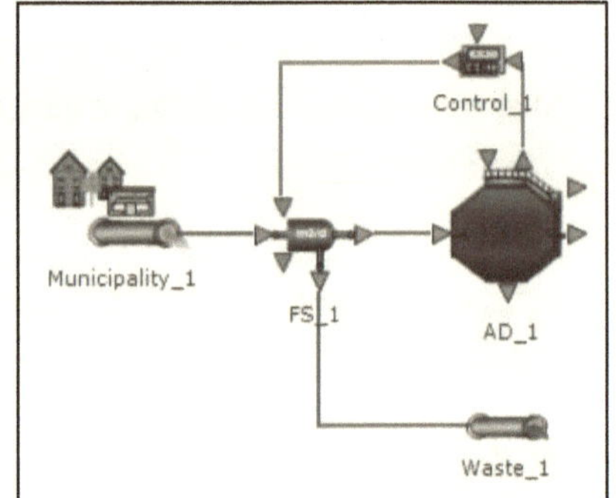

Figure 42: Anaerobic Digester with linked controller WEST ® setup

setpoint, the controller makes an appropriate compensation. The flow-through reactor was started-up towards the desired sludge age of 20 days which is established when the flow into the digester is continuous and constant at 0.7l/day from $Q_i = \frac{V}{HRT} = \frac{14}{20}$. The controller restricts the **flow into the anaerobic digester (AD_1)** based on a control parameter such as pH or the Ripley Ratio. When the flow is being diverted away from the anaerobic digester (to Waste_1), the anaerobic digester is effectively a batch reactor. The quicker the time taken for the flow to enter the anaerobic digester continuously, the shorter the time to start-up. Figure 43 shows the operation of the controller-linked anaerobic digester with the constant flow into the system. As a starting

143

point, the initial alkalinity concentration in the digester was set 250mg/L as CaCO$_3$ and a pH of 7 from the theoretical WWTP effluent with which it was filled.

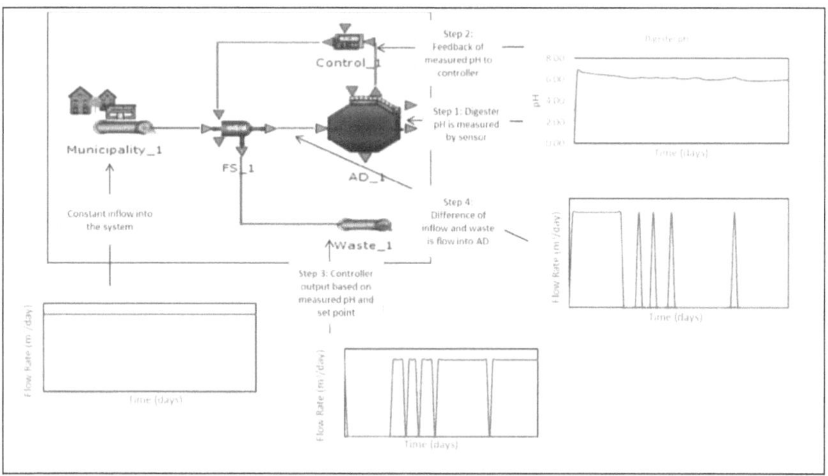

Figure 43: Diagram indicating Operation of Controller with Constant Inflow into the system

6.4.5.1 pH as the Control Parameter

The pH setpoint as the controlling parameter was varied from 6.6 to 6.2. The influent feed into the **system** for this scenario is always constant at 0.7l/day. However, the actual flow into the **digester** depends on the pH. The pH at any time step is fed back to the controller labelled "Control_1" in Figure 42 above. If the 'measured" pH in the digester is lower than the setpoint, the controller switches the flow-splitter labelled "FS_1" to divert the flow to "Waste_1" away from the digester. In theory, this should allow the digester some time to recover wherein the VFA's decrease and pH increases. When the pH rises above the setpoint, influent flow into the digester will resume once more. Flow into the digester can either be the full flow of 0.7 l/day or partial flow between 0 and 0.7l/day if the digester pH is not significantly higher than the setpoint. The term "significantly higher" is dictated by the controller factor of proportionality. If the control parameter (in this case pH) has a slow response time, too much BPO may have already entered the digester, thereby resulting in a further decrease in the pH and possible failure even when the flow entering the anaerobic digester is zero.

144

The anaerobic digester was fed with the influent listed in Table 36, and the mass of starting seed was varied for each run, starting at 1% of the final steady-state biomass concentrations. After each run, when it was seen that the pump could not stay on due to low pH, the initial seed was increased to find when a sufficient mass of acetoclastic methanogens (X_AM) allow the acetate to begin to be sufficiently consumed to keep the pH high enough so that the pump would no longer turn off. It was found that as much as 25 to 31% seed (depending on the setpoint) was required in order for the system to be run to steady state 20-day SRT. This is very costly assuming that there is no existing anaerobic digester on-site. The possible reason for this difficulty was due to this specific influent with a low pH and moderate amounts of VFA concentration as a result of the PS being partially hydrolysed in the bottom of the primary settling tank (PST). The results are summarised in Table 37.

Table 37: Summary of Results with pH as a Control Parameter

Initial % seed	Setpoint (y_S)	Able to start up?	Day on which the flow was diverted away from the digester for the first time	Day on which flow into the digester was continuous	End pH
1		No	No result available. Simulation crashed due to initial masses being too far from steady-state masses		
10		No	2	Flow into digester did not resume after day 2	4.77
15		No	2	Flow into digester did not resume after day 2	5.11
20	6.6	No	2	Flow into digester resumed on day 327 until day 375 but pH dropped once more, and flow did not enter digester again thereafter.	4.43
25		Yes	2	21	7.02
30		Yes	3	4	7.02
31		Yes	3	4	7.02
1		No	No result available. Simulation crashed due to initial masses being too far from steady-state mass		
10		No	3	Flow into digester did not resume after day 3	4.84
15	6.5	No	3	Flow into digester did not resume after day 3	5.15
20		No	3	Flow into digester did not resume after day 3	5.29
25		Yes	3	5	7.02

Initial % seed	Setpoint (y_S)	Able to start up?	Day on which the flow was diverted away from the digester for the first time	Day on which flow into the digester was continuous	End pH
30		Yes	No flow was diverted away	Flow into the digester was continuous from the start	7.02
31		Yes	No flow was diverted away	Flow into the digester was continuous from the start	7.02
1		No	No result available. Simulation crashed due to initial masses being too far from steady-state mass		
10		No	3	Flow into digester did not resume after day 3	4.76
15		No	3	Flow into digester did not resume after day 3	4.99
20	6.4	No	3	Flow into digester did not resume after day 3	5.06
25		No	3	Flow into digester did not resume after day 3	5.39
30		Yes	5	6	7.02
31		Yes	No flow was diverted away	Flow into the digester was continuous from the start	7.02
1		No	No result available. Simulation crashed due to initial masses being too far from steady-state mass		
10		No	3	Flow into digester did not resume after day 3	4.71
15		No	3	Flow into digester did not resume after day 3	4.89
20	6.3	No	3	Flow into digester did not resume after day 3	4.98
25		No	3	Flow into digester did not resume after day 3	5.06
30		No	5	Flow into digester did not resume after day 5	5.23
31		Yes	No flow was diverted away	Flow into the digester was continuous from the start	7.02
1		No	No result available. Simulation crashed due to initial masses being too far from steady-state mass		
10		No	3	Flow into digester did not resume after day 3	4.67
15		No	3	Flow into digester did not resume after day 3	4.83
20	6.2	No	3	Flow into digester did not resume after day 3	4.89
25		No	3	Flow into digester did not resume after day 3	4.97
30		No	5	Flow into digester did not resume after day 5	5.04
31		Yes	No flow was diverted away	Flow into the digester was continuous from the start	7.02

In this scenario, for the cases which could not be started-up, anaerobic digester failure occurred as a result of **overloading**, in which too much BPO had already entered the digester before the inflow into the digester could be diverted. The failure occurs because the BPO are slowly biodegradable and take some time to be hydrolysed, which in turn causes a delay in the production of the VFA's, which is what drives the pH down. By the time the BPO is hydrolysed, and the pH is low enough for the flow to be diverted, more BPO has already entered the digester which then drives the pH down even further and causes irrecoverable failure. This prevents the flow from entering the digester again, and hence, the system cannot be started up. Overloading occurs rapidly due to the large incoming load, and the system is slow to respond.

For the higher setpoint of 6.6, with 20% initial seed, the flow was diverted away from the digester for 325 days of the simulation run time, after which the pH had increased to above 6.6 and flow into the digester resumed once more for a short period in which the BPO load was substantial, thus causing a large decrease in the pH. The reason why in this case the pH increased to be high enough was due to the larger amount of seed (20%) compared with the two cases above it (10 and 15%) and also, the high setpoint which prevented too much BPO from initially entering the digester. This observation was not found with the other pH setpoints which were tested.

Because no literature information regarding the selection of the constant of proportionality was sourced, the selection of 0.5 was arbitrary. Further simulations were conducted with other proportionality constants in order to attempt to identify the most suitable one, with the setpoint remaining fixed at a pH setpoint value of 6.4. For the two other constants tested, namely 0.25 which is 50% smaller than 0.5 and 1.0 which is 50% greater than 0.5, neither could be identified as being a more suitable proportionality constant because, for both, the same amount of initial seed was required. The results are presented in Table 38.

The coding of the controller considers the proportionality constant (k_p), U_{max} and U_{min} which is the minimum and maximum control actions, u_o which is the no error action (= 0), U_{help} is a function used to compare the result of error signal and the proportionality constant to the minimum and maximum control action and y_s and y_m which is the setpoint value and sensor measured value respectively. The error and u_{help} are defined by:

$$error = y_S - y_m \tag{38}$$

$$u_{help} = u_0 + k_p \times error \tag{39}$$

An "if function" allows the u_{help} to be set to the maximum control action if u_{help} is greater than the maximum, and if u_{help} is less than the minimum, then the minimum control function will be used. For all other cases, the u_{help} will be used as the control action. This is shown by the equations below.

$$u = \begin{cases} u_{\max} & if\, u_{help} > u_{\max} \\ u_{\min} & if\, u_{help} < u_{\min} \\ u_{help} & if\, u_{\min} < u_{help} < u_{\max} \end{cases} \tag{40}$$

The control action is applied to the waste (diverted) flow, and the difference between the inflow and the control action is what flows into the digester. Therefore, with a larger k_p, the u_{help} is more likely to be greater than the maximum, resulting in the flow being diverted away from the digester.

The proportionality constant of 1, furthermore had no improvement in the time taken for the full flow to enter the anaerobic digester with 30% initial seed (took 6 days for both the 0.5 and 1 proportionality constants). This is despite that it is twice as large as 0.5 and therefore, is considered a higher gain value, requiring more power. Because no clear improved value for the constant of proportionality was identified for this case, it was investigated once more with the gradually increasing flow case (Case 3).

Table 38: Investigating the Constant of Proportionality for pH as the Control Parameter

Initial % seed	Constant of proportionality (k_p)	Able to start up?	Day on which the flow was diverted away from the digester for the first time	Day on which flow into digester was continuous	End pH
10		No	3	Flow into digester did not resume after day 3	4.76
15		No	3	Flow into digester did not resume after day 3	4.99
20	0.25	No	3	Flow into digester did not resume after day 3	5.06
25		No	3	Flow into digester did not resume after day 3	5.38
30		Yes	2	21	7.02
10		No	3	Flow into digester did not resume after day 3	4.76
15		No	3	Flow into digester did not resume after day 3	4.99
20	0.5	No	3	Flow into digester did not resume after day 3	5.06
25		No	3	Flow into digester did not resume after day 3	5.39
30		Yes	5	6	7.02
10		No	3	Flow into digester did not resume after day 3	4.76
15		No	3	Flow into digester did not resume after day 3	4.99
20	1	No	3	Flow into digester did not resume after day 3	5.06
25		No	3	Flow into digester did not resume after day 3	5.39
30		Yes	5	6	7.02

Because such a large amount of initial seed was required, another approach was investigated, i.e. the Ripley ratio (Ripley, Boyle & Converse, 1986) as a control parameter.

6.4.5.2 Ripley Ratio as the Control Parameter

For the Ripley ratio, the setpoint was initially selected based on literature values, and an exploratory investigation into the setpoint value was carried out. In well-operated digesters fed with municipal PS, the Ripley ratio, which is the ratio of the IA and PA is typically between 0.1 and 0.35 but this needs to be analysed for specific plants because high industrial contributions may change this observation for example in the case of poultry manure the anaerobic digester operated well at an IA/PA below 0.3 (Ripley, Boyle & Converse, 1986).

In the Ripley ratio, the IA, which is the net strong acid titrated (=alkalinity) between pH values of 5.7 and 4.3 and is intended to represent VFA concentration. The PA is the net strong acid titrated (=alkalinity) between the anaerobic digester sample pH and 5.7 and is intended to represent the H_2CO_3 alkalinity. While there is a reasonable correspondence between the PA and the H_2CO_3 alkalinity, the IA is a very inaccurate estimate for the VFA concentration (Ekama & Brouckaert, 2019). Because the 5-point titration of Moosbrugger et al. (1993d) provides a very accurate method for measuring both the VFA and H_2CO_3 alkalinity by simple titration of no more effort than the PA and IA, in this anaerobic digester start-up study, the Ripley ratio is defined by the actual VFA/ H_2CO_3 alkalinity concentration ratio, where both parameters are calculated by PWM_SA_AD.

With the Ripley ratio, in contrast to the pH evaluated above, a negative proportionality constant (k_p) needs to be selected since a low Ripley Ratio is desired. Therefore, if the measured Ripley Ratio is higher than the setpoint, the flow will be diverted away from the digester.

Beginning with the setpoint of 0.1 (as a lower limit setpoint from the literature) and start-up seed of 30% (because the pH-controlled system required at most 31% seed), the simulations were run. The proportionality constant was initially arbitrarily set to -0.05, but this required further investigation. The anaerobic digester started up successfully, so the percentage of seed was reduced and the effect of this analysed. The seed was reduced up to the point where the anaerobic digester could not be started up. The minimum required seed was 7%, however for this amount of initial seed, it took 1578 days (see the row shaded in grey in Table 39) for the maximum flow to enter the digester continuously, thereby attaining the required SRT of 20 days. This means that even though a constant, continuous flow of 0.7l/day was being fed into the **system**, a portion (or all of it) of it was being diverted away from the digester until day 1578. From day 1578 onwards, the full flow entered the digester constantly with nothing being diverted away to establish a steady state of 20 days SRT. This system was then tested with Ripley ratio setpoints of 0.15, 0.2, 0.25 and 0.3 while maintaining the same proportionality constant. Table 39 summarises the results.

Table 39: Results of Start-up with Ripley Ratio for Varying Setpoints

Initial % seed	Setpoint (y)	Able to start up?	Day on which the flow was diverted away from the digester for the first time	Day on which flow into the digester was continuous	End Ripley ratio value
7		No	1	Flow into the digester did not resume after day 1	0.114
10	0.1	Yes	1	110	0.012
15		Yes	1	38	0.012
20		Yes	1	20	0.012
30		Yes	2	8	0.012
7		Yes	1	1578	0.012
10		Yes	1	76	0.012
15	0.15	Yes	1	28	0.012
20		Yes	1	17	0.012
30		Yes	2	7	0.012
7		Yes	1	445	0.012
10		Yes	1	68	0.012
15	0.2	Yes	1	25	0.012
20		Yes	2	13	0.012
30		Yes	3	5	0.012
7		No	1	Flow into the digester did not resume after day 1	8.215
10	0.25	No	1	Flow into the digester did not resume after day 1	7.486
15		Yes	1	43	0.012
20		Yes	2	13	0.012
30		Yes	3	4	0.012

Initial % seed	Setpoint (y)	Able to start up?	Day on which the flow was diverted away from the digester for the first time	Day on which flow into the digester was continuous	End Ripley ratio value
7		No	1	Flow into the digester did not resume after day 1	17.69
10	0.3	No	1	Flow into the digester did not resume after day 1	19.93
15		No	1	Flow into the digester did not resume after day 1	16.38
20		Yes	2	16	0.012
30		Yes	3	1	0.012

The results show that while as little as 7% seed can be used for start-up, a 20-day SRT steady state (full flow entering the digester) takes a long time to be established. However, the start-up is also possible with 15% seed and continuous influent flows into the digester by the 25th day with a setpoint of 0.2. This is in sharp contrast to the previous case wherein pH was used as the control parameter that required as much as 31% initial seed. The higher set point of 0.3 is not conservative (safe) enough, and it is evident that it is too high to allow start-up even with the seed of 15%. Therefore, while the literature does suggest that some plants may still operate with Ripley Ratios as high as 0.35, for the purposes of this study, 0.3 is too high and results in failure due to overloading. The reason is probably due to the different definition of the Ripley ratio used in this investigation as opposed to the IA/PA definition.

The overloading is as a result of the slow rate of the BPO consumption during its' hydrolysis thereby delaying the response of the Ripley ratio and so too much BPO enters into the digester because the setpoint is not low enough which causes an accumulation of VFA even after the flow has been diverted away from the anaerobic digester. For this case, where the setpoint is 0.3 with 7% initial seed, the final acetate concentration was 503 g/m^3 and H_2CO_3 Alkalinity of 28.4g/m^3 as $CaCO_3$.

In the case where the setpoint is 0.1, a start-up with 7% seed was not possible due to the starvation of the AM rather than overloading. The low setpoint, which is too restrictive, does not allow sufficient flow into the anaerobic digester which deprives the AM of the substrate, thereby causing starvation. This was evident by the low concentration of acetate (42 g/m^3) and higher H_2CO_3 Alkalinity (363.9g/m^3 as $CaCO_3$)

compared with acetate of 503 g/m^3 and H_2CO_3 Alkalinity of 28.4g/m^3 as $CaCO_3$ with the setpoint of 0.3 (the case where failure was due to overloading). So, in effect, with the setpoint of 0.1, the anaerobic digester has not failed and gone "sour" but rather, it has failed to start-up. So, overloading is characterised by high VFA's and low H_2CO_3 Alkalinities while starvation is indicated by low VFA's and higher H_2CO_3 Alkalinities.

However, uncertainty remained regarding the magnitude of the proportionality constant of the controller. In order to investigate this, the setpoint was selected to be 0.2 because this allowed start-up in a reasonable time with all the seed amounts tested. The proportionality constant varied from -0.01 to -0.25. From the literature, larger gains result in better control but at the same requires more power (Reddi, 2011). So, in effect, one should choose the lowest possible gain (absolute value of the proportionality constant) while still attaining the requirements of a start-up with the least possible seed. However, the details of this should be investigated by Control Engineers wherein the process control is optimised with a range of practical considerations. Nevertheless, the preliminary results from these simulations are summarised in Table 40. It is evident that the largest absolute value results in the quickest start-up time. The -0.01 constant does not allow start-up for the lower seed of 7% and 10% while the constant of -0.25 only gives a minimal improvement to establish a 20-day steady state SRT compared to the -0.05 constant. Therefore, the selection of -0.05 seemed reasonable for the case with Ripley Ratio as the control parameter.

Table 40: Investigating the Constant of Proportionality for the Ripley Ratio as the Control Parameter

Initial % seed	Proportionality constant (k_p)	Able to start up?	Day on which flow to the digester was diverted away for the first time	Day on which full flow entered digester continuously	End Ripley ratio value
7		No	1	Flow into the digester did not resume after day 1	8.201
10	-0.01	No	1	Flow into the digester did not resume after day 1	6.751
15		Yes	1	32	0.012
20		Yes	2	13	0.012
30		Yes	2	5	0.012
7	-0.05	Yes	1	445	0.012

Initial % seed	Proportionality constant (k_p)	Able to start up?	Day on which flow to the digester was diverted away for the first time	Day on which full flow entered digester continuously	End Ripley ratio value
10		Yes	1	68	0.012
15		Yes	1	25	0.012
20		Yes	2	13	0.012
30		Yes	3	5	0.012
7		Yes	1	410	0.012
10		Yes	1	67	0.012
15	-0.25	Yes	1	25	0.012
20		Yes	2	13	0.012
30		Yes	2	5	0.012

6.4.6 Digester Start-up with Proportionally Increasing Flow and Without a Controller (Case 2)

In this start-up investigation, the anaerobic digester was fed with the flow which increases by a fixed percentage each day while the concentrations in the influent remained constant, thereby resulting in constantly increasing influent flux proportional to the flow rate. For a shorter time to start-up, the percentage increase in the influent flow rate at each day is higher than for a longer start-up time. The daily increase in influent flow in terms of the start-up duration is given by $t^{\frac{1}{t}} - 1$ where t is the start-up duration in days.

This system was not linked to a controller. The flow pattern was determined by setting the time to start-up and consequently determining the required percentage increase in influent flow in order to attain maximum flow at a set time for a start-up. By increasing the time required to start-up (full flow), the required seed was decreased. Table 41 below summarises the results. As can be seen, for a seed of 11%, the quickest duration to start up is 600 days and 100 days to start up requires 14% seed. 100 days is a long time to start up, but it requires significantly less seed compared with the controller defined influent flow scenarios discussed above.

Days to max flow	% initial seed of total volume required
1100	10%
1000	11%
900	11%
800	11%
700	11%
600	11%
500	12%
400	12%
300	12%
200	12%
100	13%
50	16%
30	18%

For the number of days until the maximum flow is attained as listed in Table 41, the corresponding initial seed amounts reflect the minimum required seed in order for the anaerobic digester to not fail and hence start-up successfully. For any initial seed less than the amount listed for each scenario, the anaerobic digester fails to start up due to irrecoverable digester failure. For example, with 1100 days to maximum flow, and if 10% seed is initially used, the BPO and acidogen biomass initially is low due to the slow rate of the incoming BPO. As the influent flow gradually increases, the BPO begins to increase as does the acidogen biomass. The acidogen biomass increases slowly due to the slow BPO hydrolysis which produces glucose slowly as a substrate for the acidogen. During the acidogen utilisation of glucose, acetate, hydrogen and propionate are formed. There is a peak in the acetate (S_VFA) concentration until the acetoclastic methanogens are able to grow sufficiently fast to cope with the acetate load. Failure is prevented because the amount of initial seed allowed the acetoclastic methanogens to successfully grow sufficiently fast to avoid VFA accumulation and acetoclastic methanogen inhibition by low pH. Figure 44 shows these observations.

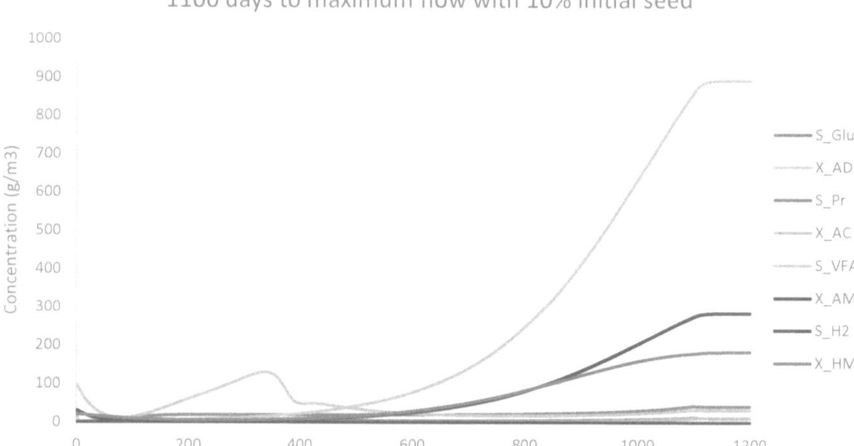

Figure 44: Graph of Concentrations in AD for 1100 days to Maximum Flow with 10% initial seed

If 9% of the initial seed is used instead of 10%, the system cannot start up, and anaerobic digester failure occurs. Although the acetate concentration and acetoclastic methanogen biomass concentrations are very similar in both cases up until before day 250, the minuscule difference actually causes the failure because the small amount of additional acetate in the case where 9% seed is used, causes a slightly lower pH which causes a slower growth, i.e. the tipping point to irrecoverable failure is reached. So, as more acetate is formed, the amount not being utilised by the acetoclastic methanogens due to their inhibition eventually accumulates to the point where the acetoclastic methanogens are fully inhibited. It is at this point where there is a sharp increase in the acetate. Figure 45 compares the acetate profiles for the 9% and 10% initial biomass seed cases, and Figure 46 compares the acetoclastic methanogen biomass concentrations.

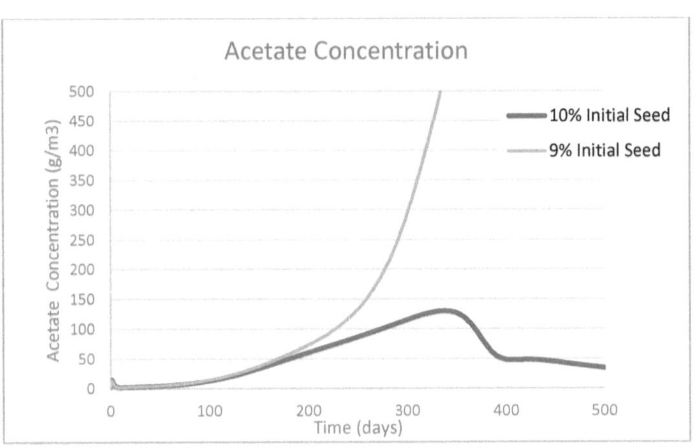

Figure 45: Acetate Concentration in the Anaerobic Digester for 9% and 10% Initial Seed

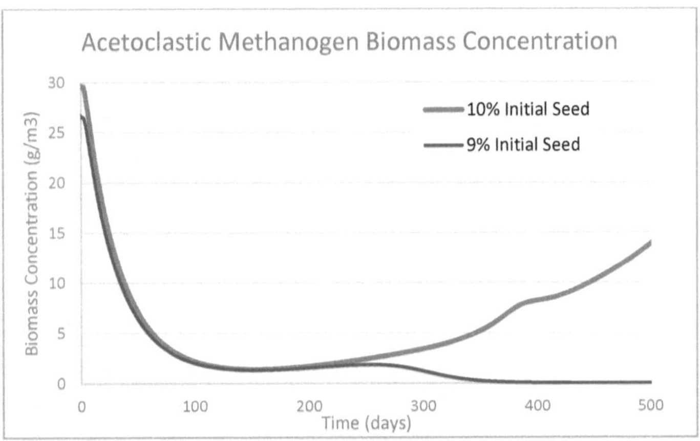

Figure 46: Acetoclastic Methanogen Concentration in the Anaerobic Digester for 9 and 10% Initial Seed

For the above 9% and 10% seed cases, initially, the acetoclastic methanogens die off over the first approximately 100 days due to the slow incoming BPO and the slow rate at which the acetoclastic methanogens grow. This is illustrated in Figure 46. Therefore, when the amount of acetate is sufficient for the methanogens to begin to grow, a significant amount of the acetoclastic methanogens have already died off due to starvation. This is despite the significant VFA concentration in the influent. So, while the actual failure for 9% seed at a maximum flow of 1100 days is due to the amount of

157

acetate causing pH inhibition which prevents it from being utilised, the initial trigger for failure was due to initial starvation of the methanogens which slowed their overall growth rate. This is perhaps due to using Monod kinetics for the maximum specific growth rate of the acetoclastic methanogens which depends on the bulk liquid acetate concentration. So, a question that arises here is "is it appropriate to use Monod kinetics for the growth processes which follow hydrolysis, i.e. acidogenesis, acetogenesis, acetoclastic methanogenesis and hydrogenotrophic methanogenesis?". Saturation kinetics may, in fact, be more appropriate for digester start-up and anaerobic digester failure. This will require further investigation.

When the maximum flow into the digester is set to 30 days, the required seed is larger (18%). However, when the seed is lower than 18%, the mode of failure differs from the above 9% seed case. In this case, because of the larger amount of initial biomass seed, it results in a greater overall growth rate for the acetoclastic methanogens. In addition, a higher acetate load also serves to increase the growth rate, and hence the acetoclastic methanogen biomass does not undergo the initial degree of die-off as it had with the 9% and 10% seed cases above where the maximum flow was attained after 1100 days. For 17% initial seed, failure occurs at the point where the high accumulation of acetate results in a low pH and subsequent inhibition of the methanogens. In a time of three days, the acetate concentration had more than tripled, resulting in complete inhibition of the acetoclastic methanogens. So, failure, in this case, was due to the acetate load overwhelming the acetoclastic methanogens.

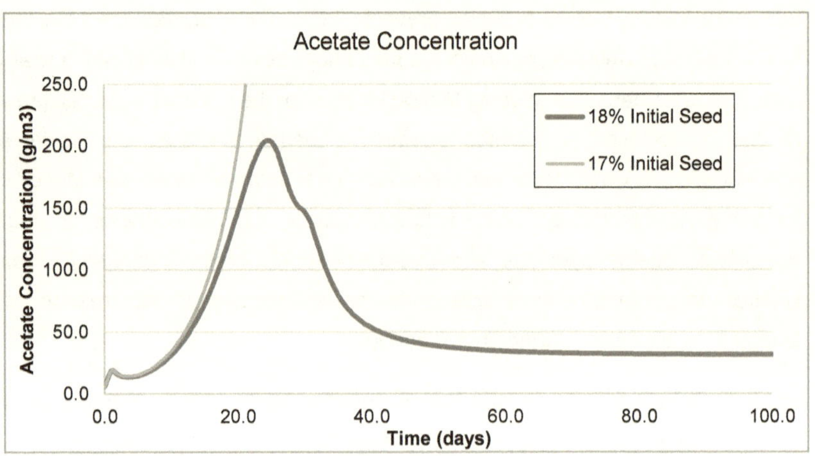

Figure 47:Acetate Concentration for 17% and 18% Initial seed

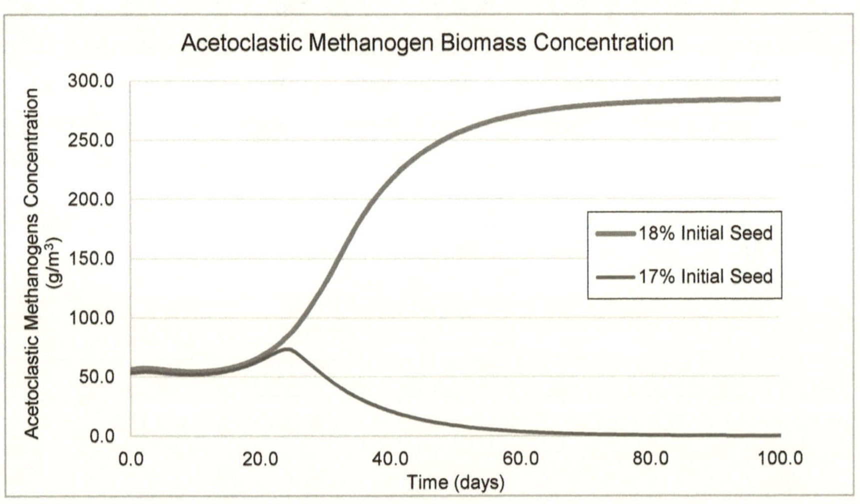

Figure 48: Acetoclastic Methanogen Biomass Concentration for 17% and 18% Initial Seed

The most probable cause for the lengthy time to start up is that the model had been calibrated to a specific UASB dataset which was not under start-up conditions. Nevertheless, because the UASB system induces anaerobic digester failure at the

159

bottom of the bed, the model is a better representation of reality under anaerobic digester failure conditions. Prior to calibration, the model would be unable to predict a start-up scenario because the acetate never increased due to its utilisation rate being too high. Such a model cannot be used in an actual digester start-up because digesters are prone to failure during start-up if overloaded with too large amounts of BPO.

Because the proportional influent flow applies the BPO load on the system via the gradually increasing flow, the digester is given sufficient time to grow acidogenic and acetoclastic methanogen biomass such that when the load on the digester is larger, there is sufficient biomass to cope with the load. At the same time, starvation does not occur because there is a constant influent load coming in, which serves to grow the biomass. This raises the possibility of a digester start-up scenario wherein a gradually increasing flow is fed to the digester, while at the same time being linked to a controller with the possibility of decreasing the time to start-up and/or the required seed thereby making the system more efficient.

An aspect of controlled systems is that the system run time with a controller should be as fast or faster than an uncontrolled system for the value of the controller to be realised. Running the digester start-up with the proportional influent flow but without a controller provides a baseline for comparison with the proportional but controlled inflow, which is discussed below.

6.4.7 Proportionally Increasing Flow Linked with a Controller (Case 3)

6.4.7.1 pH as the Control Parameter

a) Maximum Flow into the System after 100 days

Varying initial seed and setpoints

The proportionally increasing uncontrolled inflow system could be started up in 100 days with 13% seed. Thus, this flow pattern was now fed to a system with a pH controller to compare if the start-up could be achieved with less seed over the same time to start-up. The pH setpoint value dictates how soon the influent flow is diverted away from the digester. The daily increase in flow is specified by the time duration to attain the maximum flow. The proportionally increasing and later maximum flow into the system is not necessarily the flow that enters the digester. Once the user-defined inflow has been specified, there is a closed-loop action which occurs. The closed-loop action measures the digester pH using a pH sensor. This information is feedback to the controller, which compares the pH setpoint value to the measured pH in order to determine an error signal. Depending on the error signal, the control action is to either waste (divert) no flow, waste all the flow or waste a portion of the flow entering the system. The difference between the user-defined inflow and the waste flow is the flow that enters the digester. Figure 49 shows these closed-loop controller steps as Step 1 to Step 4. The user-defined proportional influent flow enters the system regardless of what is occurring with the sensor and the controller. For the specific case in Figure 49, the flow does not enter the anaerobic digester (except for the first few days and a few spikes thereafter) because the pH is below the setpoint, nearly all the time. From the diagram, it is clear that the controller actions the waste flow rather than the flow into the anaerobic digester.

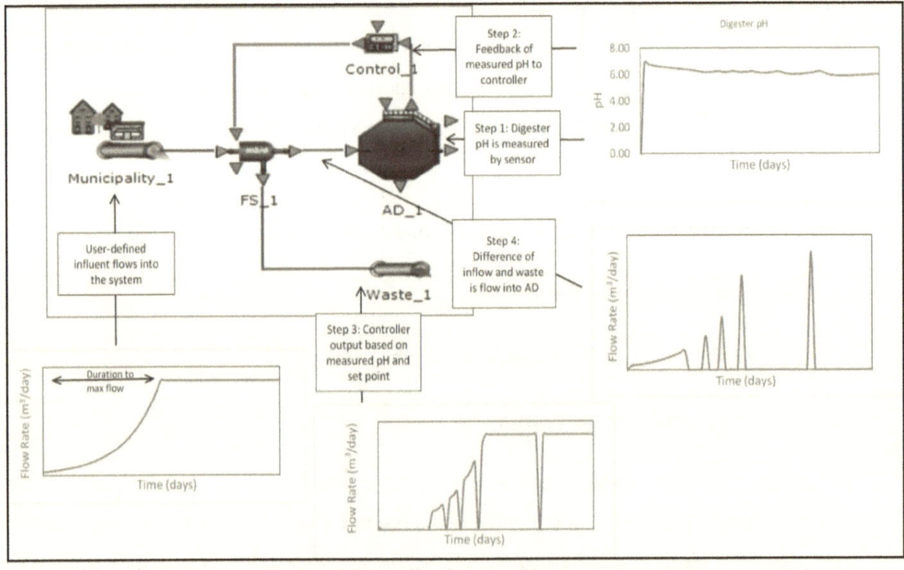

Figure 49: Explanatory Diagram of Controlled Variable Inflow System

The results of the simulations are summarised in Table 42. Failure to start up is characterised either as starvation, starvation followed by overloading or overloading.

Descriptions regarding these different modes of failure to start up are discussed after the table. The shaded rows are the scenarios which were able to start up successfully.

Table 42: Results of a Start-up with a Linked Controller with the Maximum Flow at 100 days

% Initial Seed	pH setpoint (y)	Able to start-up?	Day on which the flow into the digester was diverted away for the first time	Day on which full flow entered digester (continuously)	Reason for non-start-up (if applicable)
10		No	3	Flow into the digester did not resume after day 3	Starvation
11	6.6	No	3	Flow into the digester did not resume after day 3	Starvation
12		No	3	Flow into the digester did not resume after day 3	Starvation
13		No	3	Flow into the digester did not resume after day 3	Starvation
10		No	9	Flow into the digester did not resume after day 9	Starvation
11	6.5	No	9	Flow into the digester did not resume after day 9	Starvation
12		No	10	Flow into the digester did not resume after day 10	Starvation
13		No	10	Flow into the digester did not resume after day 10	Starvation
10		No	20	The full flow never enters the digester	Starvation
11	6.4	No	23	The full flow never enters the digester	Starvation followed by overload
12		No	26	The full flow never enters the digester	Starvation followed by overload
13		Yes	29	100	-
10		No	26	The full flow never enters the digester	Overload
11	6.3	Yes	35	107	-
12		Yes	38	100	-
13		Yes	No flow was diverted away	100	-
10	6.2	No	34	The full flow never enters the digester	Overload

% Initial Seed	pH setpoint (y)	Able to start-up?	Day on which the flow into the digester was diverted away for the first time	Day on which full flow entered digester (continuously)	Reason for non-start-up (if applicable)
11		No	38	The full flow never enters the digester	Overload
12		No	44	The full flow never enters the digester	Overload
13		Yes	No flow was diverted away	100	-

As can be seen, the scenario with the setpoint at 6.3 can be started up with 11% seed which is 2% less seed required compared with not using a controller, although the maximum flow enters the anaerobic digester at day 107 instead of day 100. In the cases where no flow was diverted away from the digester, the controller is serving no purpose. This was for the case where the initial seed amount is greater than or equal to the amount of seed required for the uncontrolled system (i.e. 13% for 100 days to the maximum flow into the system).

The cases which could not be started up were either due to starvation, starvation followed by overloading, or by overloading only. In the case of starvation, the pH setpoint value is too high to allow flow into the digester. The methanogens are unable to grow and begin to die, which is indicated by their low concentration. However, the pH is just slightly below the setpoint, and H_2CO_3 alkalinity has not decreased very much.

For the cases of starvation followed by overloading, this is due to initially low quantities of flow entering the digester, which does not allow the methanogens to grow sufficiently. This causes starvation. As the pH and flow rate increase, the acetate load on the anaerobic digester increases once more for a short time. When this occurs, though the flow into the anaerobic digester is brief, too much BPO enters the anaerobic digester due to the delay in the response time of the pH as a result of the slow hydrolysis rate. This causes a sharp decrease in the pH and H_2CO_3 alkalinity.

Lastly, for the case where overloading occurred, this was as a result of too low a pH setpoint value to prevent too much BPO from entering the anaerobic digester. This causes an immediate high accumulation of acetate, a decrease in H_2CO_3 alkalinity and a substantial decrease in pH.

An important observation is that while literature often indicates that the methanogens are primarily inactive below a pH of 6.6, this does not justify setting the setpoint to 6.6 because the selection of the setpoint is very much plant-specific. Due to this low pH influent, a setpoint which is too high causes starvation of the organisms. This observation shows that investigations into the plant behaviour prior to developing a control philosophy is necessary.

A different way to analyse the above table would be to consider the results from Table 41, wherein the system was not linked with a controller. As can be seen, 11% seed for the system without a controller requires 600 days to start-up whereas in the case linked with a controller, this is achieved in 107 days. So, the controlled system allows the maximum flow into the digester to be attained quicker than the uncontrolled system thereby showing its worth.

An investigation into the setpoint effect on start-up outcome

Uncertainty still existed as to why 11% and 12% initial seed allows start-up with a set point at 6.3 but not with any of the other tested pH setpoint values (Table 42). Intuitively one would think that setting a higher setpoint could prevent failure better. But a system that is too restrictive does not allow sufficient substrate to enter the system and the organisms begin to starve. In order to investigate this aspect further, the specific AM growth rate (r_{AM}/X_{AM}) was plotted against time for three different scenarios with a constant seed of 12%. This was the net growth rate accounting for the death of biomass via the decay process and the inhibition due to pH. The first one with a setpoint of 6.2, then 6.3 and lastly 6.4. The maximum specific growth rate of the acetoclastic methanogens at a pH of 7.0 is 0.495/day. When the specific growth rate is below 0 /day, it indicates that the death rate is greater than the growth rate. With a pH setpoint of 6.4, the specific growth rate was barely above zero, indicating that the biomass was experiencing decay for most of the simulation. This is as a result of starvation caused by a limited substrate.

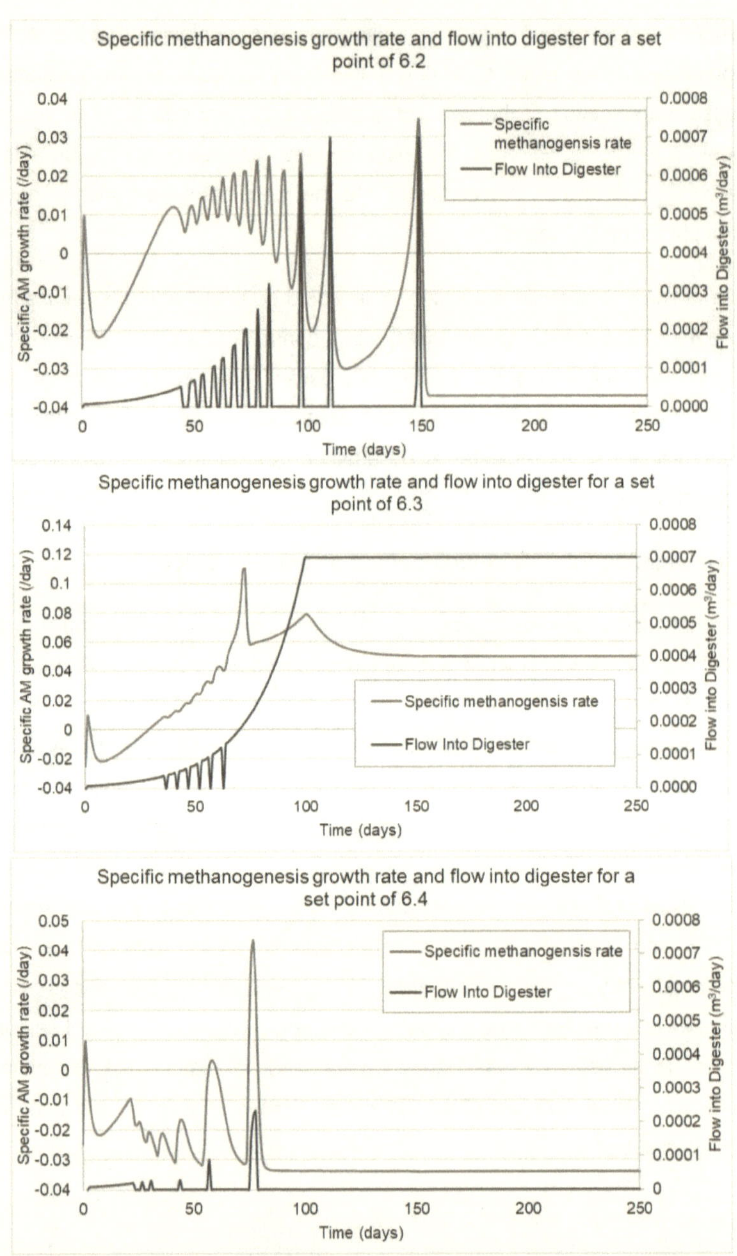

Figure 50: Graphs Indicating the effect of the Setpoint on Inflow into the Digester and the Specific Acetoclastic Methanogenesis Rate

For the case when the pH setpoint value was set at 6.4, the system was found to be too restrictive. Thus, the system could not start-up because of insufficient feed which caused initial starvation of the methanogens early on. This was evident when considering that at day 30, the concentration of acetoclastic methanogen biomass was $20g/m^3$ and $10g/m^3$ of acetate for a setpoint of 6.4, while for pH setpoint values of 6.2 had an acetoclastic methanogen biomass concentration of $25g/m^3$ and acetate concentration of $75g/m^3$. Although the pH had increased slightly to allow flow into the anaerobic digester for the case of a setpoint of 6.4, the biomass concentration was already too low to utilise the substrate and thus the substrate accumulated, dropping the pH.

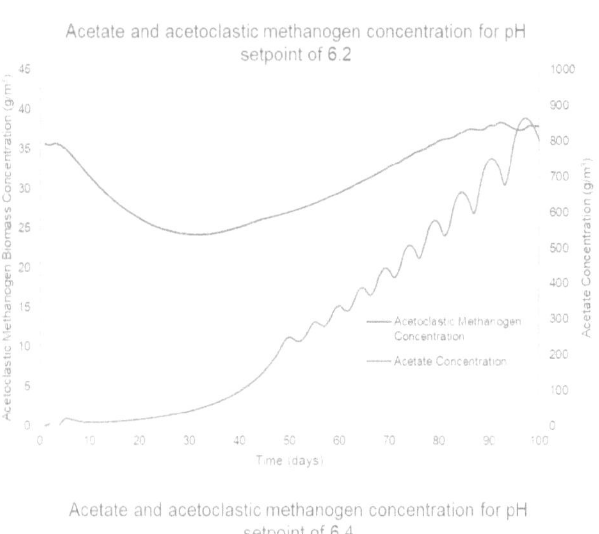

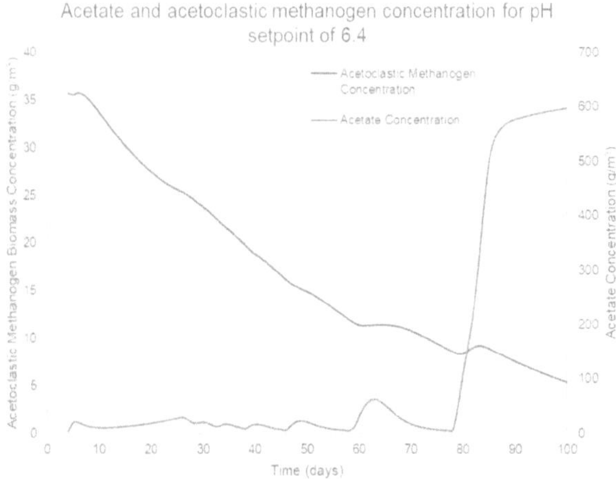

Figure 51: Acetate and acetoclastic methanogen biomass concentration for pH setpoints of 6.2 and 6.4

In the case of the setpoint at 6.2, the system performed identically to the system with a setpoint of 6.3 until day 36 when the pumped turned off for the first time for the system with a setpoint of 6.3. The pump turned off for the first time on day 44 with a pH setpoint value of 6.2. This is due to this system having a bigger range over which it operates. However, the operation range is too large, and thus, the system gets overloaded too easily. Too much acetate accumulates in the system, thereby lowering the pH and preventing start-up.

Interestingly, the setpoint controls how the system fails during start-up. For pH setpoint values from 6.4 upwards (with 12% seed or less), the system fails due to the starvation of the organisms because insufficient feed enters the digester. If the mode of failure is purely starvation, there exists a large amount of H_2CO_3 alkalinity and a pH which is just below the setpoint value. For pH setpoints of 6.2 or less, the system fails during start-up because the setpoint values are not conservative (safe) and restrictive enough. This failure mode is characterised by low pH values, low H_2CO_3 alkalinity and high acetate values (i.e. overloading). This observation may be specific to this study because the influent acetate concentrations are high, pH values are low so setting a high pH setpoint means that the system cannot run due to the low incoming pH. From the point of view of assessing how effective using a controller is, it can be said that the use of a controller results in a lower seed requirement for start-up in the same time compared with a system which does not use a controller.

An investigation into the effect of the proportionality constant (k_p)

The effect of the proportionality constant (k_p) was then evaluated once more with proportionally increasing flow because the constant flow scenario provided no insight into the effect of the proportionality constant. The proportionality constant was varied in order to assess its effect on the system wherein the maximum flow was attained at 100 days. In all the initial runs (with pH as the controlling parameter) described above, k_p was set at 0.5. For testing the effect of k_p, the setpoint selected was 6.4 with 12% initial seed, which failed to start-up due to starvation, as described in section 6.4.7.1 above. The simulations were run at k_p values of 0.25, 0.2 and 0.192. The results are shown in Table 43. Intuitively, one would have thought that preventing flow to the anaerobic digester (higher k_p) allows the digester to perform better, but this generally causes starvation. This was surprising considering the high acetate concentration and further investigations are required to evaluate the suitability of Monod kinetics and if its use for the growth process is the cause of the starvation.

For the case where k_p = 0.192, the system can start up successfully while start-up was not possible for the other k_p values evaluated. With a lower k_p value (i.e. 0.192), the flow enters the digester over a broader range of pH values (because the u_{help} function is more likely to be less than the maximum control action) and so the flow is diverted away from the digester later and flow into the digester is resumed sooner which allows the methanogens to be fed with substrate thereby preventing failure. The graph in Figure 52 below shows this phenomenon until day 80.

While the case with k_p = 0.192 allowed successful start-up due to it operating over a broader range of pH values, any value lower than this causes failure due to overloading as was the case with k_p = 0.1.. Similarly, always allowing flow into the anaerobic digester with 12% initial seed (maximum flow into the system at 100 days) would result in overloading as was evident from Case 2 tested above (no linked controller scenario). Thus, with the use of a controller, and a suitable setpoint and factor of proportionality, the digester can start-up successfully with less seed than the system with no linked controller. So, while an observed pattern is evident regarding the factor of proportionality, it is not possible to decide on an actual "correct" value because of the lack of measured start-up data.

Table 43: Effect of the Proportionality Constant on Start-up

k_p	pH setpoint (y)	Day on which the flow into the digester was diverted away for the first time	Day on which the full flow entered the digester (continuously)	end pH
0.1		27	The full flow never enters the digester	4.922
0.192		21	876	6.901
0.2	6.4 12% seed	21	The full flow never enters the digester	6.034
0.25		21	The full flow never enters the digester	5.888
0.5		21	The full flow never enters the digester	5.728

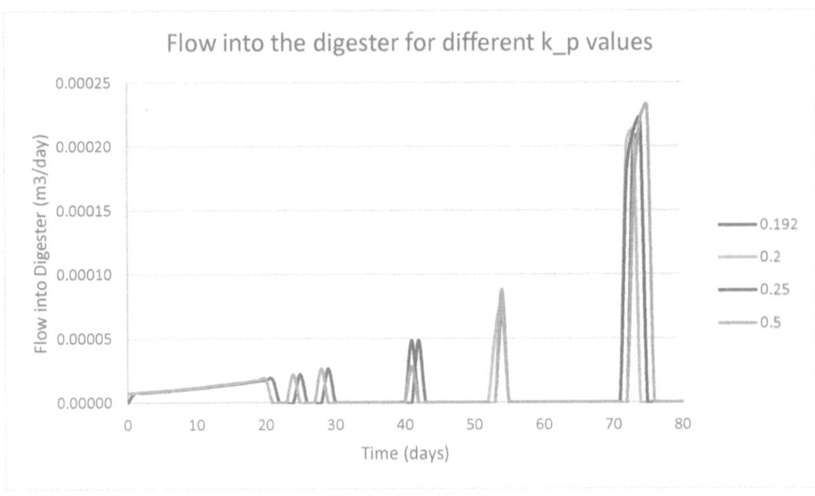

Figure 52: Graph Indicating the Effect of the Proportionality Constant

The system responded well to the pump turning off when the pH drops too low, but it fares better if the feed resumes soon because a delay in resuming the feed actually causes digester failure via biomass starvation (Figure 46, Figure 47 and Figure 50). The controller-linked digester can start-up successfully in the same amount of time as the uncontrolled proportionally increasing flow-fed digester with less seed, so the benefit of a controller is clearly evident. Stopping the feed for too long results in starvation of the biomass or when the flow resumes, it is too high of a load for the biomass because part of them died off.

b)   Full Flow into the System at 30 days

Thus far for all the Case 3 (proportionally increasing flow with a linked controller) results described above, the proportionally increasing flow results in the maximum flow into the system being attained at 100 days. Because the full flow into the system is only attained at 100 days for the scenarios investigated above, the quickest time for the full flow to enter the digester will never be quicker than that, i.e. less than 100 days. This is a long time in practice. Due to this, the pH setpoint analysis was repeated with the maximum flow into the system being attained at 30 days. Although the maximum flow into the

system is at 30 days, the full flow into the **digester** may take longer depending on the flow which is diverted away from the digester initially.

If the pH measured by the sensor is higher than the setpoint, the flow will enter the digester. However, the time at which the **maximum** flow enters the digester continuously may be different from the time which the maximum flow enters the system set by the user-defined influent. The four graphs in Figure 53 show this phenomenon. While the full flow enters the system at day 30, the full flow enters the digester continuously only from day 62 because the pH was mostly below the setpoint for most of the time between day 8 and 55. Table 44 summarises the results and the shaded rows represent the scenarios which can be started up successfully.

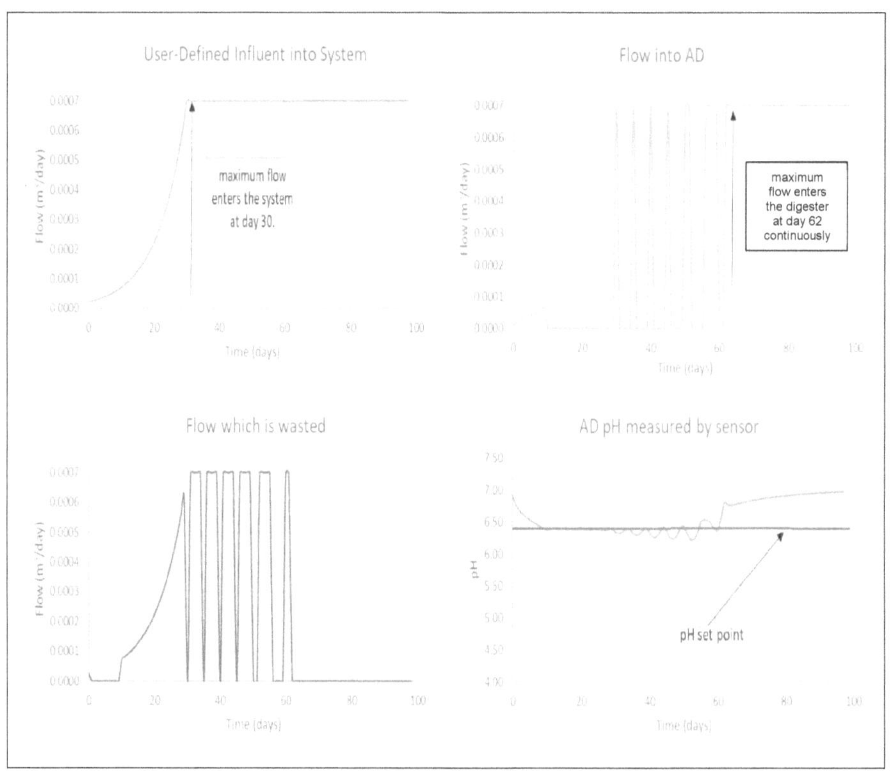

Table 44: Results of a Start-up with a Linked Controller with the Maximum Flow at 30 days

% Initial Seed	pH setpoint (y)	Able to start-up?	Day on which the flow was diverted away from the digester for the first time	Day on which full flow entered digester (continuously)	Reason for non-start-up (if applicable)
13	6.6	No	4	The full flow never enters the digester	Starvation
14		No	4	The full flow never enters the digester	Starvation
15		No	4	The full flow never enters the digester	Starvation
16		No	5	The full flow never enters the digester	Starvation
17		No	5	The full flow never enters the digester	Starvation
18		No	5	The full flow never enters the digester	Starvation
13	6.5	No	6	The full flow never enters the digester	Starvation
14		No	7	The full flow never enters the digester	Starvation
15		No	8	The full flow never enters the digester	Starvation
16		No	8	The full flow never enters the digester	Starvation followed by overloading
17		Yes	9	58	-
18		Yes	10	53	-
12	6.4	No	10	The full flow never enters the digester	Starvation followed by overloading
13		Yes	10	62	-
14		Yes	10	45	-
15		Yes	12	35	-
16		Yes	13	34	-
17		Yes	16	31	-
18		Yes	No flow was diverted away	30	-
13	6.3	No	12	The full flow never enters the digester	Overloading
14		Yes	14	47	-
15		Yes	16	61	-
16		Yes	18	58	-
17		Yes	22	54	-
18		Yes	No flow was diverted away	30	-

13	6.2	No		14	The full flow never enters the digester	Overloading
14		No		16	The full flow never enters the digester	Overloading
15		No		17	The full flow never enters the digester	Overloading
16		No		19	The full flow never enters the digester	Overloading
17		No		23	The full flow never enters the digester	Overloading
18		Yes	No flow was diverted away	30		-

Consider from Table 44 that with a setpoint of 6.4, the system can start-up with 13% initial seed. The results from Table 41 were used as a comparison to identify if the controlled system operates quicker than the uncontrolled system. While the uncontrolled system requires at least 18% initial seed for successful start-up in 30 days (see Table 41), which is 5% more seed than the amount required for the system linked with a controller, the system linked with a controller has the maximum flow entering the digester only at day 62 (see Table 44). In the uncontrolled system, however, 100 days is required in order to successfully start-up a system wherein the seed is 13%. Thus, in terms of time, the controller-linked system starts up 38 days quicker (100 days for the uncontrolled system compared with 62 days for the controlled system). Therefore, the benefit of the controller-linked system is clearly evident.

The difference between the scenario wherein full flow into the system was attained at 30 days compared to that of the 100 days, is that in the 100-day case, the flow and hence the substrate load on the system per day (initially before the maximum flow is attained) are very much lower than for the 30-day case. This results in a significantly greater concentration of acetate which is not utilised for the 30-day case. For the setpoint of 6.4, with 12% seed, the acetate concentration was $2672g/m^3$ for the 100-day case, but it was $5384g/m^3$ in the 30-day case. So, while it is preferable that the start-up is complete in the shortest time possible, overloading is a very likely possibility when the load is very high.

6.4.7.2 Ripley Ratio as the Control Parameter

When the Ripley ratio was used as the control parameter with **constant inflow** (section 6.4.5.2) into the system, 7% initial seed was required for successful start-up (although the flow into the digester was only at the maximum continuously from day 445 - see Table 39). From the gradually increasing flow tested previously with no linked controller, Table 41 indicates that for 30 days to start up, 18% seed was required. Thus, combining the proportionally increasing flow combined with a Ripley ratio controller may allow either less initial seed or quicker time to start-up successfully. Because the Ripley Ratio allowed start-up with just 7% seed for a constant inflow (albeit lengthy time taken for the full flow to enter the anaerobic digester), the system with a controller was now fed with gradually increasing flow such that maximum flow into the **system** was attained at 30 days. This was done so to determine if the Ripley Ratio controller can provide a quicker start-up time with less initial seed compared with the pH as the control parameter. The selected setpoint was 0.2, with a constant of proportionality of -0.05. This was because these combinations of parameters allowed start-up over the biggest range of initial seed amounts tested.

Table 45: Summary of a Start-up with Ripley Ratio as a Control Parameter for Gradually Increasing Flow

% Initial Seed	Able to start up?	Days operated until pump switches off the first time	Day on which full flow entered digester (continuously)
6%	No	3	The full flow never entered the digester
7%	Yes	4	374
8%	Yes	5	149
9%	Yes	6	98
10%	Yes	7	73
11%	Yes	8	59
12%	Yes	9	50
13%	Yes	the pump did not switch off	43
14%	Yes	the pump did not switch off	38
15%	Yes	the pump did not switch off	35
16%	Yes	the pump did not switch off	32
17%	Yes	the pump did not switch off	30
18%	Yes	the pump did not switch off	30

From Table 45, a clear pattern is evident in that as the percentage of seed is increased, the day on which the full flow enters the digester occurs earlier. For 6% initial seed, the start-up was not possible because the initial seed amount is too low to allow the pump to remain on beyond day 3. Although the start-up is possible with 7%, a long time is required before the full flow enters the digester (374 days). However, in contrast to constant inflow with the Ripley Ratio as the control parameter (as reflected in Table 39), where the full flow is attained at 445 days, this is a slight improvement. Considering that the system can be started up with 7% seed in not much of a longer duration, the use of gradually increasing flow, in this case, is not necessarily warranted. When comparing the Ripley Ratio to the pH as the control parameter; however, the benefit of the Ripley Ratio is evident because start-up is attained with significantly less seed and generally shorter duration. For initial seed amounts between 13 and 18%, the pump runs continuously from the beginning of the simulation, and thus, the controller serves no purpose.

6.4.8 Evaluation of the Monod Kinetics for the Acetoclastic Methanogen Growth Process

Generally, when dynamic models are employed, they are fed with the results of a steady-state simulation which allows the bulk liquid concentrations of substrates to be in a proportional amount to the biomass concentration which is mediating the process. However, in digester start-up scenarios, where the initial seed in the digester is low, but the bulk liquid substrate concentrations are high, the Monod kinetics, which relies on the bulk liquid concentrations, predict starvation of the biomass when one would expect them to grow at maximum rate due to the availability of substrate. This phenomenon was also observed in activated sludge models where predictions of aerated lagoons were required. With the use of Monod kinetics, the biomass struggled to grow. This was however not observed if saturation kinetics were used instead. This poses the question of whether or not the utilisation of Monod kinetics is appropriate for start-up conditions and would saturation kinetics improve the modelling outcomes. However, with the lack of available data, it becomes challenging to know which kinetic formula is more superior.

Thus, instead of exploring the possible use of saturation kinetics with limited available data, the Monod specific growth rate to the maximum specific growth rate ratio of the acetoclastic methanogens was plotted against time for each of the start-up scenarios described in Sections 6.4.5 to 6.4.7 above. The acetoclastic methanogens were already identified to be the critical organism group responsible for maintaining the balance between the organism groups. For this investigation, the specific growth rate did not include the inhibition terms nor the decay of the biomass due to their death. Thus, this focused only on the suitability of the growth rate term.

The ratios highlighted a clear pattern for either overloading (which had a high ratio), starvation (which had a low ratio) and successful start-up which was in between the two. For all cases analysed, if overloading was the mode of failure, the ratio was greater than 0.89. This shows that the reaction rate is not being slowed down by an imbalance of biomass and substrate. For these overloaded cases, the use of saturation kinetics would not improve the modelling outcomes, because the biomass is not starving due to limited substrate. In contrast, the ratio for starvation was less than 0.05, indicating that the biomass growth was restricted. For these cases of starvation, saturation kinetics may help especially because literature often cites the pH lower limit of the methanogens to be at 6.6 whereas at this setpoint value, starvation occurs and a low specific growth rate to the maximum growth rate of acetoclastic methanogens is observed.

In Case 1, where the system was fed with a constant flow with a linked controller, for the cases **where start-up was not possible**, the failure mode was always overloading and the value of the ratio at steady state was greater than 0.89. Failure due to starvation did not occur for this case because the full incoming load enters the system allowing the organism groups to be sufficiently fed. Figure 54 shows the observed ratio as a function of time during the simulation under overloading conditions.

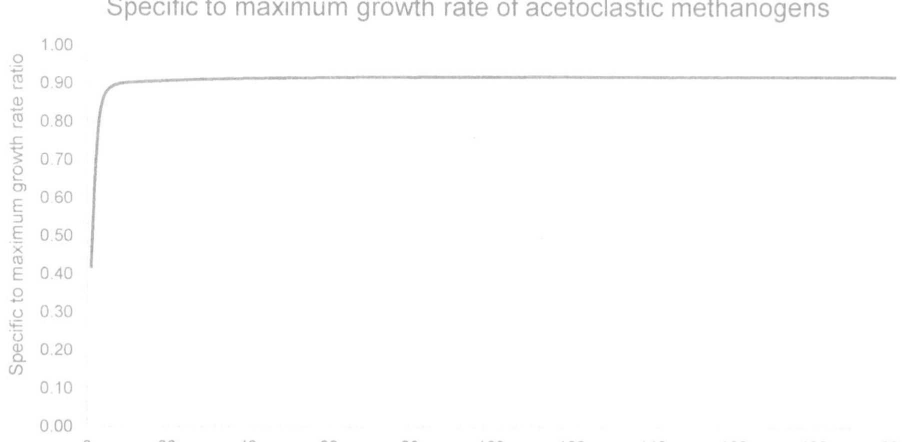

Figure 54: Specific Growth Rate to the Maximum Specific Growth Rate Ratio of AM in Case 1 under Overloading Conditions

For Case 3, where failure to start up could either occur due to overloading or starvation, the ratios vary between being very low (0.05) for cases of starvation or very high (greater than 0.89) for cases of overloading. Literature frequently cites the AM's as being significantly inhibited for pH values less than 6.6 and that these low pH values can cause irrecoverable failure. However, for this case, for setpoint values of 6.5 and 6.6, the failure to start up was caused by starvation. Thus, the rate of the reaction is limited and **may** be resolved using saturation kinetics. Figure 55 shows the observed pattern for the specific growth rate to the maximum specific growth rate ratio for starvation conditions.

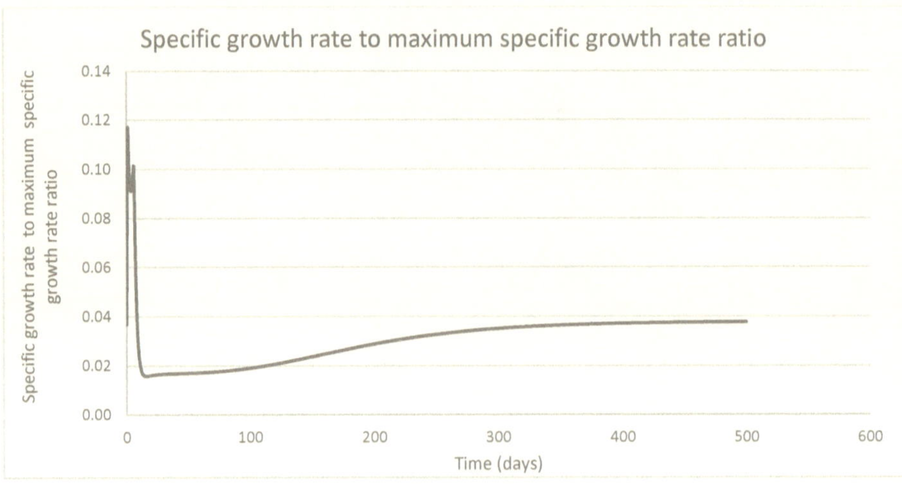

Figure 55: Specific Growth Rate to the Maximum Specific Growth Rate Ratio of AM for starvation conditions

The difficulty is that without actual start-up data, determining whether Monod and saturation kinetics are better for modelling the start-up is not possible. This illustrates that for conditions wherein Monod kinetics predicts starvation, saturation kinetics may predict start-up better.

The graphs below compare the start-up results for setpoint values between 6.2 to 6.5. This was for Case 3, wherein the proportionally increasing flow was fed to a system with a linked controller. The maximum flow into the system occurred at day 100 and 12% initial seed was used. At the lower limit of 6.2, failure to start-up is as a result of overloading. Successful start-up occurs at a setpoint of 6.3, while starvation occurs at a set point of 6.5. For the setpoint of 6.4, the system initially undergoes starvation followed by overloading.

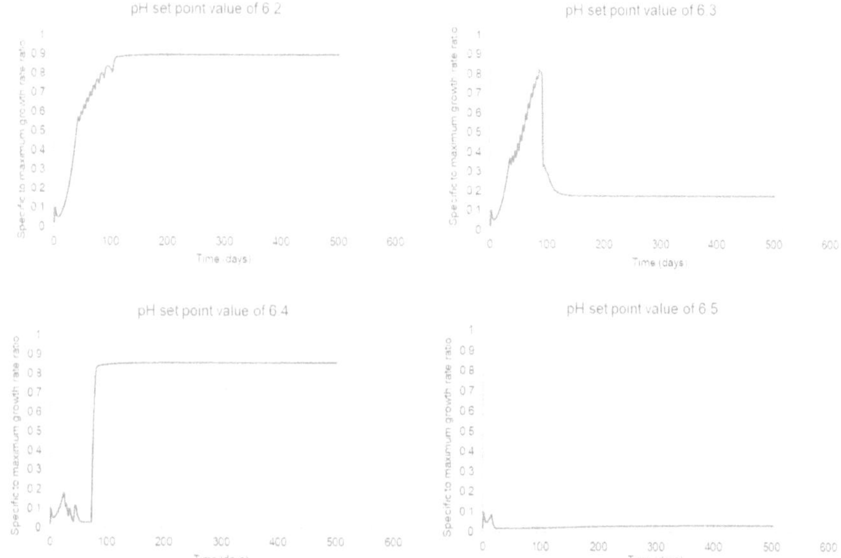

Figure 56: The Specific Growth Rate to Maximum Specific Growth Rate Ratio Graphs for Setpoint Values between 6.2 and 6.6

6.4.9 Effect of Feeding Unhydrolyzed Sludge and Glucose

Another aspect which was considered was to analyse how the model predicts feeding unhydrolysed sludge. Because the Izzett (1992) dataset had a high concentration of acetate and a low pH, it was assumed that this sludge represents partially hydrolysed sludge. In order to adjust the sludge to conditions prior to the assumed partial hydrolysis, the COD of the influent acetate was added to the COD of the influent BPO, the influent pH was raised to 7.5, and the influent alkalinity was increased to 250mg/L as $CaCO_3$.

This influent was fed to a digester at constant flow and load with 11% initial seed. For the system with partially hydrolysed sludge, the start-up was achieved within 1000 days. However, with the unhydrolysed sludge, the system could only start-up with feeding at such a slow rate that full flow was only attained at 3000 days. The initial digester seed was then changed to 14%. Whereas the first system with partially hydrolysed sludge could start-up within 100 days, in this case, the system with unhydrolysed sludge took 300 days to start-up. The acetoclastic methanogens, begin to die off because they are not being "fed" any acetate, such that when the acetate is present following the

178

hydrolysis of the primary sludge, there are insufficient methanogens to deal with the accumulating acetate.

In the first scenario, where the assumed partially hydrolysed sludge was fed, there is already a presence of acetate in the influent. Thus, initially, it allows the methanogens to grow to a certain extent. So, when the load on the methanogens becomes higher following the hydrolysis of the particulate organics (which were not hydrolysed as yet) the methanogens are able to cope. So, partially hydrolysing the sludge allows failure to be more easily prevented.

A final task was to compare the digester start-up procedure if glucose was used as the feed instead of BPO. The concentration of BPO was converted into glucose by considering the COD of the particulate organics. For the same system feeding BPO that started up in 100 days with 14% seed, the system which was fed glucose required an initial seed of 50%. This difference can be attributed to the fact that the BPO contain latent alkalinity, which is transferred to the aqueous phase, thereby raising the pH and preventing the accumulation of acetate due to a low pH. This latent alkalinity is in the form of organically bound nitrogen which gets released as ammonia (NH_3), and which reacts with dissolved H_2CO_3 to form NH_4^+ and HCO_3^-. In addition, glucose acidifies more rapidly thereby placing a much larger load on the acetoclastic methanogens. The PWM_SA_AD model predicts what would be expected from an anaerobic digester under these different conditions, confirming that it is suitably calibrated.

6.5 CONCLUSION

The PWM_SA_AD model was calibrated in this study and subsequently used in different applications within this chapter. After calibration, the model proved to be useful to predict an anaerobic digester failure scenario which was attained by decreasing the influent alkalinity until failure occurred.

The minimum required alkalinity set a baseline for comparison with failure due to the sensitivity of acetogens to hydrogen. When testing these conditions, it was found that allowing the acetogens to be more sensitive to hydrogen serves to cushion the anaerobic digester against failure because the acetate load from glucose utilisation occurs prior to the acetate load from propionate utilisation. However, propionate accumulation was found not to be the cause of failure, and for the cases simulated, the failure was always caused by the acetate accumulation instead. With the completion of this testing, it was found that the PWM_SA_AD model was a reasonably good representation of reality and had the capability to be used in control operations to explore how anaerobic digester failure could be prevented.

A comparison was also drawn between PWM_SA_AD and ADM1 in terms of comparing the way in which the models predicted anaerobic digester failure. This was done by programming the ADM1 bioprocesses into PWM_SA_AD and using a switching function to select which model was used. With ADM1, failure occurs at higher influent alkalinity due to a slower acetoclastic methanogen specific growth rate, greater sensitivity of propionate to hydrogen and possibly because of the higher utilisation rate of glucose, which places a greater load on the acetoclastic methanogens.

Lastly, the calibrated model was applied to anaerobic digester start-up conditions using a primary sludge dataset of Izzett (1992). The hydrolysis rate was modelled with saturation kinetics, while the four AD bioprocesses that follow hydrolysis were modelled with Monod kinetics. Even though there is a delay in the response time of system parameters such as pH and the Ripley ratio, due to the slow rate of the BPO hydrolysis, they were selected as the controlling parameters because pH and Ripley ratio measurements are easily obtainable directly with a probe and the 5-point titration method, respectively. The feeding of the influent at a constant flow and load requires either a long time to start up or large amounts of initial seed. This was evident in the system which used the Ripley ratio as the control parameter in that although it could be started up with as little as 7% seed, the flow into the anaerobic digester was continuous only from day 1578. The system which used the pH as the control parameter could not be started up unless the initial seed was 25% or greater. Thus, the problem was identified to be with the constant flow and load on these systems and this required further optimisation.

As a means to optimise the start-up further, the flow and load could be gradually increased, allowing the time taken for the anaerobic digester to operate continuously, at the desired SRT of 20 days, to be decreased relative to the constant flow and load scenario. So, to provide a comparison, the gradually increasing influent was fed directly to the anaerobic digester without a linked controller. This showed that for the maximum flow to enter the anaerobic digester at 100 days, 14% initial seed was required, and

19% initial seed was required for the maximum flow to enter the anaerobic digester at 30 days. Although not directly comparable to the constant flow scenario linked with a controller, the system with gradually increasing flow allows the maximum flow to enter the system continuously with less seed. It was envisaged that linking a controller would allow the seed or time taken for the maximum flow to enter the system to be decreased further, and so this was done. With pH as the control parameter, 11% seed was required for the scenario where the maximum flow into the system was attained at 100 days (although it took 107 days for the maximum flow to enter the anaerobic digester) and 13% seed was required for the case where the maximum flow into the system was attained at 30 days (with the maximum flow into the anaerobic digester occurring at day 62). With the Ripley ratio as the control parameter, the maximum flow into the anaerobic digester occurred at day 374 although only 7% seed was required.

A major outcome of this start-up investigation showed that the model predicts that organisms begin to starve where setpoints are too conservative, due to flow restriction into the system. Plotting the specific growth rate to the maximum specific growth rate ratio showed that kinetically, the specific growth rate was too slow due to low bulk liquid concentration of acetate. This may be due to a limitation with Monod kinetics, and further investigations are required to evaluate whether the use of saturation kinetics for the four anaerobic digester processes following hydrolysis may improve modelling outcomes under the extreme conditions of anaerobic digester failure.

Finally, using partially hydrolysed sludge allows the acetoclastic methanogens to begin growing slowly from the beginning of the simulation such that when the load on them has increased, they are able to cope with the higher concentration of acetate. In contrast, by using unhydrolyzed sludge, the methanogens concentration immediately begins to decrease as they starve during the hydrolysis process awaiting their substrate of acetate. It follows that the utilisation of partially hydrolysed sludge allows either less initial seed to be used or shorter duration to start-up and provides greater stability in the anaerobic digester. Glucose, as an alternative to hydrolysed sludge, acidifies at a rapid rate and places a tremendous load on the acetoclastic methanogens. Thus, systems which are fed with glucose, although the control parameters do not experience the slow response time as in the case of feeding BPO to the system, resulting in rapid overloading conditions and irrecoverable failure.

The value of the completed calibration exercise is clearly evident because, without the lower maximum specific growth rate for the acetoclastic methanogens, an accumulation of the acetate would not have occurred. This meant that the model prior to calibration could not be used in anaerobic digester failure or start-up behaviour evaluation because it was not an accurate representation of reality.

7 CONCLUSIONS AND RECOMMENDATIONS

7.1 CONCLUSIONS

This study developed an AD model, which is suitably calibrated to model AD dynamics such as impending digester failure and digester start-up. This is required because, while AD has numerous benefits, including low energy requirements and low sludge yield, it is notorious for being susceptible to failure and is regarded as unstable and sensitive, and thus not operated at optimal levels. An improved, calibrated AD model is helpful to enhance understanding of AD dynamics during failure. This, in turn, will allow for anaerobic digesters to be operated closer to capacity.

Sewage sludge AD models to date have been calibrated only for the hydrolysis step, which is rate-limiting under normal conditions. This is suitable for applications wherein municipal sewage sludge is digested, and there are no strongly varying flows and loads into the digester. However, under conditions approaching failure, or during start-up, the processes following hydrolysis may become the rate-limiting ones. Through these processes, intermediate products such as acetate, propionate and hydrogen accumulate. Calibration of the hydrolysis step alone is therefore no longer adequate, and calibration of the remaining processes (acidogenesis, acetogenesis, acetoclastic methanogenesis and hydrogenotrophic methanogenesis) is also required. The most appropriate model to do this was investigated.

The PWM_SA wastewater treatment plant model is a three-phase plantwide model in WEST® which includes mineral precipitation and gas evolution. With an external speciation routine, this model separates the faster aqueous reactions from the slower physical (gas exchange and precipitation) and biological processes thereby eliminating model stiffness. Furthermore, PWM_SA characterizes the organics' composition with routine wastewater treatment measurements, making it more practical than other AD models such as ADM1 which uses carbohydrates, lipids and proteins to characterise the influent. Being a subset of PWM_SA, the AD model PWM_SA_AD has all the features of the plant-wide model and the versatility of the modelling platform WEST®. This AD model was thus selected as the most suitable base model for AD calibration in order to make developments which would allow modelling digester failure and start-up.

A detailed stoichiometric and kinetic verification of the model was completed to confirm the model's accuracy and elemental mass balances. This brought about greater confidence and reliability in the model. Though the PWM_SA_AD model is based on the UCTSDM, the manner in which the equations are derived differ. The addition of appropriate conversion terms, however, allowed the models to be exactly aligned.

PWM_SA_AD was calibrated against the UASB system. This AD system was selected because temporary failure takes place in the lower level of the bed, exhibited by an accumulation of intermediate substrates (acetate, propionate and by implication hydrogen). This required PWM_SA_AD to be able to model different bed levels of the UASB so that this temporary failure at the bottom of the bed could be accurately modelled. This was done by coding six in-series, completely-mixed, anaerobic digesters into WEST®, with each digester capable of retaining solids based on a solids retention factor. In running the six-in-series UASB system, instability arose where the initial AD biomasses in the digesters were too different from the final steady-state biomasses. This problem was overcome with the use of steady state spreadsheet models to calculate close to final initial steady-state masses so that the integrated reactor configuration could be run to generate meaningful results for the calibration procedure.

Within the calibration protocol, one of the primary concerns is generally the identification of the most important parameters that require calibration. Multiple simulations allowed a large dataset to be generated. This dataset was used to fit linear models varying in complexity and detail. Within these linear models, the importance of the primary explanatories on the model was evaluated by considering the β coefficients. In the more complex linear models, which included interaction or secondary explanatories, the lasso method allowed for the identification of the most important explanatories. The Morris screening method then served as confirmation of those explanatories identified as important by the lasso method. The outcome of the sensitivity analysis indicated the retention factors and Monod maximum specific growth rates and half-saturation coefficients as the most important parameters to calibrate in the AD model.

A major challenge with the sensitivity analysis arises because the variables' concentrations are very much associated with other variables' concentrations. In addition, they include a degree of causality, meaning that one variable can cause another variable to respond in a certain way. However, in typical statistical methods employed, as well as those used in this study, there is no way to decipher between causality and association merely by considering the β coefficients. This illustrates that while significant emphasis is placed on a statistical sensitivity analysis, the intricacies in the system have to be understood from a bioprocess perspective as well, which is where the steady-state models proved to be very useful.

During the calibration procedure, despite a good prediction for the intermediate AD products (acetate and propionate) over the bed height, the pH was found to be poorly predicted by the model. In order to correct this, the equilibrium assumption between the headspace CO_2 partial pressure and aqueous phase CO_2 concentration was revisited. The gas evolution, which was the remaining variable to affect the pH, was changed to rate-controlled process driven by a mass transfer rate. This was considered to be more accurate because, in the lower levels of the UASB, the CO_2 gas is not in equilibrium with the headspace. With this correction, the predicted pH matched the observed pH well along the height of the UASB. By virtue of its calibration to a UASB system, wherein AD failure takes place in the lower part of the bed, the model was deemed capable of predicting anaerobic digester failure and start-up conditions. Prior to calibration, the AD model did not predict an accumulation of intermediates, which cause AD failure because the maximum specific growth rates were too high. With specific growth rates calibrated to UASB datasets, however, the AD model is more suitable to be applied to real anaerobic digesters.

The calibrated set of constants is not necessarily unique and may not be universally applicable. This is because the actual residual glucose and bed solids concentrations were not measured in the UASB datasets. Therefore, a maximum specific growth rate of the acidogens had to be assumed, on which the solids retention factor of the first anaerobic digester in the UASB series of six depends. Once these two parameters are set, the maximum specific growth rates of the other three AD organism groups are no longer arbitrary. Had the residual glucose and bed VSS solids concentration been measured in the UASB dataset, this degree of freedom of the maximum specific growth rate of the acidogens and the solids retention factor of the first anaerobic digester would have existed.

The calibrated AD model was applied to analyse how the AD system performs under complete and irrecoverable failure conditions. This was done by gradually decreasing the influent alkalinity to the UASB system from the dataset value of 6000 mg/L as $CaCO_3$. Irrecoverable failure (high effluent intermediate concentrations) occurred at influent alkalinity of 4200mg/L as $CaCO_3$. This was the point at which the specific growth rate of the acetoclastic methanogens fell below the minimum rate required to utilize the high acetate concentration. Lower influent alkalinity, such as the abovementioned 4200mg/L as $CaCO_3$, results in lower buffer capacity thereby reducing the bed pH, causing an inhibition of the acetoclastic methanogens below a critical rate.

One of the commonly thought causes of AD failure in the literature is acetogen inhibition to hydrogen. From these simulations, it was found that the load on the acetoclastic methanogens is, in fact, less when the acetogens are inhibited and not producing acetate and, thus, to an extent, forestall failure. The acetogens were made more sensitive to hydrogen, which inhibits them at a lower hydrogen concentration, and thereafter the influent alkalinity was gradually reduced. Now irrecoverable failure occurred at influent alkalinity of 4000mg/L as $CaCO_3$, which verified the theory of the effect of the acetogens sensitivity to hydrogen. Thus, with the acetogens more inhibited by hydrogen, the system fails at lower influent alkalinity, of 4000mg/L as $CaCO_3$, compared with the less inhibited acetogens.

The above modes of UASB failure predicted by PWM_SA_AD were compared with ADM1, to draw a comparison between how the two models differ in their prediction of failure. In order to maintain consistency in the pH prediction, ADM1 was coded into PWM_SA_AD as an independent subset using the same external speciation routine. One of the documented apparent shortfalls of ADM1 is its inability to predict AD failure. However, it was found that ADM1 does indeed predict failure in the same way as PWM_SA_AD, but it occurred at higher influent alkalinity (5000mg/L as $CaCO_3$) than PWM_SA_AD. This was mainly because the specific growth rate of the acetoclastic methanogens in ADM1 is slower than in the PWM_SA_AD model calibrated to the UASB data. With the same biomass kinetic constants, ADM1 predicts AD failure similarly as PWM_SA_AD.

In order to get an adequate understanding of how the model behaves under digester start-up conditions, which is essentially avoiding AD failure by not overloading it too much too soon, three different start-up cases were investigated (with Case 1 and Case 3 having two subparts each). These cases were (1) setting the influent pump at the final steady-state flow rate to establish the required HRT (=20d) but switching off and on a controller which could either be a pH controller or a Ripley ratio controller, (2) daily increasing the influent flow by a fixed proportion of the final steady-state flow ($t^{1/t}$-1), where t is the start-up duration in days, (3) same as (2) but adding a pH controller or a Ripley ratio controller. These cases used the calibrated kinetic constants and were compared against the digester start-up using the default kinetic constant prior to calibration. In all these simulations, typical primary sewage sludge was fed to the AD containing a percentage of seed and filled with wastewater treatment plant effluent with 250 mg/L as $CaCO_3$ alkalinity. The percentage of seed is the percentage of the mass of sludge in the AD at the final steady state with which the AD is filled at start-up and contains the same proportions of the four AD biomass groups. In the simulations, the duration (days) to a successful start-up for different starting percentage of seed was determined.

For the constant flow case with pH as the control parameter (Case 1) at least 25% initial seed was required in order to start up the digester successfully, but it started up immediately without any oscillation of the pump turning on or off. This is impractical due to the cost factor of transporting, from an off-site location, large amounts of AD sludge to be used as seed. This is in sharp contrast to the start-up predictions using the uncalibrated kinetic constants, which allowed start-up with just 1% initial seed. So, although 25% initial seed is a high amount, it reflects the associated challenges with digester start-up better than the model prior to calibration.

The problem of using pH as the control parameter is its slow response time. So, the control parameter was changed to the Ripley ratio, while maintaining the constant flow into the system. Although with the Ripley ratio as the control parameter the digester started-up with as little as 7% initial seed, the influent flow into the digester is only continuous from day 1578 - requiring over 4 years to start-up an AD is not practical at all. Although the Ripley ratio, therefore, allowed start-up with less seed and had a better response time than the pH, the constant (yet switched on and off) influent flow places a large load on the system, making the start-up duration longer. Interestingly, these issues were not encountered at all with the kinetic constants prior to calibration, indicating that the kinetic rates were too high.

Further simulations were done to find a start-up method that requires less seed sludge than the case with a pH controller and a shorter start-up duration than with a Ripley ratio controller. In this case, where the flow into the digester was constant (but switched on and off), it was found that the digester failed to start-up due to BPO overload. Because the hydrolysis rate of BPO is slow, too much BPO enters the AD to avert failure due to low pH acetoclastic methanogen inhibition (the BPO overload). Further investigations were carried out to find a way to prevent this.

Because of the failure to start-up being a result of BPO overload, the system was fed with a proportionally increasing flow (Case 2) to determine whether anaerobic digester start-up could be attained with less seed and quicker than required for the constant flow scenarios. As a baseline for comparison, initially, this proportionally increasing flow was fed into the digester without a controller. The daily increase of inflow per day is ($t^{1/t} - 1$) and is set by the duration for start-up t days. Interestingly, if the full flow enters the digester after 1100 days, due to the slowly increasing influent flow, the acetoclastic methanogens biomass decreases due to endogenous respiration and never recover, a starvation failure. This is in sharp contrast to when the full flow is attained in 30 days, in which the acetoclastic methanogens become overloaded due to BPO overload if the initial seed is less than 18%. So, the failure to start-up can be either due to BPO overload, as was the case for Case 1 and proportionally increasing flow over less than 30 days to full flow (within Case 2 but quick duration to attain the maximum flow), or acetoclastic methanogen starvation, for proportionally increasing flow over more than 1100 days (Case 2 with a long duration to attain the maximum flow). For Case 2, between 30 and 1100 days start up, the higher the percentage of seed, the faster the start-up duration. The occurrence of acetoclastic methanogen starvation was surprising because the primary sludge feed contained a high VFA concentration (~2200 mg/L) and that should have allowed sufficient acetoclastic methanogen growth. The reason for the starvation despite the high amount of available substrate required further investigation.

The further investigations of the starvation of the acetoclastic methanogens showed that when plotting a graph of the specific growth rate to the maximum specific growth rate ratio of the acetoclastic methanogens under starvation conditions, the ratio is extremely low. The ratio is low because the bulk liquid acetate concentration is low from Monod kinetics. The limitations of Monod kinetics are apparent here because the specific growth rate of the acetoclastic methanogens under these conditions is expected to be high. The use of saturation kinetics will maintain a close to maximum specific growth rates when the substrate concentration is low and may prevent this type of failure in AD modelling. Whether or not this will produce a more realistic AD model requires further investigation.

Lastly, the proportionally increasing flow start-up (Case 3) was linked with a pH controller or a Ripley ratio controller. For the system with a linked pH controller, the full flow into the digester to achieve the 20d HRT occurred at day 62 with 13% initial seed, in contrast to 100 days with the system not linked with a controller. With the Ripley ratio as the control parameter, 10% seed allows the full flow to enter the digester at day 73, compared with 1100 days required for the system with no linked controller. So, the Ripley ratio as the control parameter greatly influences the duration to start-up. Even with as little as 7% seed start-up is possible, although it takes just over a year (374 days) to do so. The benefit of the controller in both the cases with pH as the control parameter and the Ripley ratio as the control parameter is clearly evident as the start-up is possible in a shorter duration than the uncontrolled system.

The selection of the setpoint greatly dictated the failure mechanism. With a pH setpoint which is too low, failure to start-up was as a result of BPO overload which resulted in a high acetate concentration and a low pH thereby slowing the acetoclastic methanogens below the tipping point to start-up. This mode of failure could also be due to a low percentage of seed or due to the slow hydrolysis and acidification of BPO. In contrast, acetoclastic methanogen starvation occurred as a result of a pH setpoint which was too high, thereby preventing flow from entering the digester and depriving the organisms of the substrate. Similar observations were made with the Ripley ratio as the control parameter wherein starvation occurred if the setpoint was low (too conservative) and overloading occurred if the setpoint was too high (not sufficiently conservative).

As evidenced in this study the system prior to calibration posed no difficulty in attaining start-up. This was despite the low influent pH and high influent acetate concentration. In addition, acetate never accumulated, and the pH did not drop low enough to cause any inhibition, albeit only having an initial seed of 1%. Thus, the kinetic constants prior to calibration were much too high for the AD model to exhibit AD failure. In contrast, the calibrated model exhibited AD failure as observed in the bottom of a UASB and, therefore, was considered to represent better the challenges associated with AD start-up. Some of the AD model responses during start-up cases, such as the effect of the initial seed required and the pH which frequently decreased below 6.6 to 6.2 which strongly inhibits the acetoclastic methanogens, are similar to observations in practice. From this, it is clear that the model has shown greater capabilities in predicting digester start-up, which the previous model failed to do.

Through the findings discussed above, the model calibration carried out in this book advances AD modelling in a way that makes the prediction of AD failure and digester start-up possible. This has not been possible with previous models which were not suitably calibrated.

7.2 RECOMMENDATIONS

Although the dataset selected for this study for calibration is the one with the most available data, one of the primary difficulties encountered during this study was the lack of measurements in the dataset. The accumulation of the intermediate products allowed the calibration of the organism groups following hydrolysis, but the lack of measurements, such as methane gas flow, sludge wastage and sludge bed solids concentrations, prevented a COD balance from being calculated over the experimental results. Had these measurements been available, the degree of freedom in the selection of the maximum specific growth rate of the acidogens and, consequently, the retention factor of the first anaerobic digester would not exist. This would place greater reliance on the calibrated results. The principles and outcomes discovered through this study are, however, still relevant and provides valuable insight into the modes of digester failure and start-up.

Regarding the sensitivity analysis, many researchers have placed great emphasis on completing a sensitivity analysis prior to calibration. The results from the sensitivity analysis highlight the important parameters that require calibration. It also indicates the non-influential parameters which can be set to the literature values and do not require calibration. However, the accuracy of the sensitivity analysis greatly depends on the degree to which the linear models derived for the variables explain the variance in the data (adjusted R^2). When there is a low adjusted R^2, the linear model accounts for a low proportion of the variance in the data. The low adjusted R^2 could be as a result of a high degree of interaction in the data between the parameters and/or variables. However, without the assistance of specialist statisticians, it becomes difficult and complex to improve the correlation, resulting in a more complex model. Within these more complex models, although they may explain the variance in the data indicated by higher adjusted R^2, it may be difficult to draw inferences about the important parameters. In essence, although the statistical sensitivity analysis was meant to simplify the output, its value is lost due to its complexity. As a result, the complexity becomes one of trying to understand statistics instead of bioprocesses.

Steady-state models, wherein the time-derivative is set to 0, allows the user to clearly identify the most important parameters simply by looking at and analysing the resultant equation. The equation may not be a linear one, but considering the extreme scenarios allows one to determine whether the resulting concentrations significantly depend on the maximum specific growth rates, half-saturation coefficients, yield values or death rates. Thus, the benefit of steady-state models is that it provides insight into the operational outcomes of the dynamic models without the complex interactivity. Steady-state models, in combination with the statistical sensitivity analysis, therefore provides greater insight into important parameters, rather than just relying on the sensitivity analysis alone. This combined approach is therefore recommended for future studies wherein calibration is required.

Some literature indicates that digester failure prediction can only occur if the hydrogen partial pressure effect is included in both the acidogenesis as well as acetogenesis steps. This was however proven not to be true in this study. Although PWM_SA_AD includes the effect of high hydrogen partial pressure on both acidogenesis and acetogenesis, and ADM1 only includes the effect of hydrogen on acetogenesis, both models predicted an occurrence of failure as was shown in Chapter 6. ADM1 produced propionate as a standard acidogenesis product, unlike PWM_SA_AD which only produces propionate under high partial pressure of hydrogen conditions. ADM1 nevertheless predicted AD failure to occur as did PWM_SA_AD. While it cannot be concluded, at this stage, which model is better in terms of failure prediction, it is clear that both models are capable of predicting failure. More investigation into this aspect is warranted. This will be possible with more investigation into UASB temporary failure conditions with adequate measurements.

Giovannini et al. (2016) found that many authors even recommend the use of hydrogen as a warning indicator when other typical indicators may not provide early warning of gradual overload. So, although, based on literature, hydrogen has been found to be an important intermediate in preventing failure, more investigations are required to determine whether this is, in fact, true, as this was found not to be the case in this study. The mode by which hydrogen inhibition occurs is based on the unfavourable conditions prevailing which stalls the acetogenic process. However, theoretical partial pressures have been documented but not necessarily measured in lab-scale systems. This further makes it challenging to determine the required inhibiting hydrogen concentration into modelling-based systems. Because hydrogen was not measured in the operated UASBs, it becomes difficult to determine the role that hydrogen plays in these scenarios. It is therefore recommended that further developments are made regarding the role that hydrogen plays in the AD process. This will provide greater insight into the failure mode and will assist in determining whether propionate is a standard acidogenesis product or not.

Furthermore, with the aim of making the PWM_SA_AD model more widely applicable to industrial wastewater, the model requires further developments to include processes such as the Stickland reactions and possibly fatty acid oxidation. The feasibility and benefit of including these processes will be investigated in future research where deemed necessary. For more improved modelling of sulphate reduction, glucose conversion by SRB's will also be considered as an addition to the model under sulphidogen

Although an automated digester start-up control strategy was envisaged, the restrictions of the WEST® software meant that this aspect could not be investigated in sufficient detail. The further recommended research requires an investigation into start-up with varying set points being used during the start-up simulation. The use of a lower pH set point initially will allow the system to run and thus prevent starvation, and once the organisms have reached their temporary capacity, if the set point could be increased, the flow to the digester would be restricted thereby allowing the organisms to recover somewhat. This may prevent the challenges which arose regarding starvation and overloading.

Lastly, the conditions within an anaerobic digester during start-up are very far from the final equilibrium. This poses an important limitation with the use of Monod kinetics to describe the bioprocess rates. Monod kinetics are useful to describe dynamics when bioprocess systems are close to equilibrium/steady state and are pushed away from the equilibrium. This is what happens in a UASB. However, in digester start-up, this is not the case. The initial biomass concentrations are very far from their final steady-state/equilibrium concentrations, and the bulk liquid substrate concentrations are very low, resulting in a low specific growth rate. In contrast, with saturation kinetics, while both biomass and substrate to biomass concentrations are low, their ratio is high, resulting in a close to maximum specific growth rate, allowing the biomass to grow at a high rate. While the hydrolysis process in PWM_SA_AD uses saturation kinetics, further investigation into the behaviour of the Monod and saturation kinetics for modelling the other four bioprocesses will prove to be useful in order to improve modelling outcomes of digester start-up. There is, therefore, scope for further investigations to be carried out with a specific focus on adequate measurements of residual COD, gas flow, hydrogen concentration, and bed concentrations. Furthermore, more detailed bed measurements may allow more detailed biofilm modelling to be completed which may enhance modelling outcomes.

With these additional measurements, it may then be possible to obtain a global set of maximum specific growth rates for the processes which follow hydrolysis. Plant specific performance could then be reduced to BMP tests in order to quantify the organics composition which will allow calibration of the hydrolysis process.

8 REFERENCES

Ahring, B.K. 2003. *Biomethanation I*. V. 81.

Ahring, B.K., Sandberg, M. & Angelidaki, I. 1995. Volatile fatty acids as indicators of process imbalance in anaerobic digestors. *Applied Microbiology and Biotechnology*. 43:559–565.

Allison, J.D., Brown, D.S. & Novo-Gradac, K.J. 1991. *MINTEQA2/PRODEFA2, A geochemical assessment model for environmental systems: Version 3.0 Users manual*.

Angelidaki, I., Karakashev, D., Batstone, D.J., Plugge, C.M. & Stams, A.J.M. 2011. *Biomethanation and its potential*. 1st ed. V. 494. Elsevier Inc.

Batstone, D.J., Keller, J., Angelidaki, I., Kalyuzhnyi, S.V., Pavlostathis, S.G., Rozzi, A., Sanders, W.T., Siegrist, H., et al. 2002. The IWA Anaerobic Digestion Model No 1 (ADM1). *Water Science and Technology*. 45(10):65–73.

Bhuiyan, M.I.H., Mavinic, D.S. & Beckie, R.D. 2009. Determination of temperature dependence of electrical conductivity and its relationship with ionic strength of anaerobic digester supernatant, for struvite formation. *Journal of Environmental Engineering*. 135(11):1221–1226.

Billing, A. & Dold, P. 1988. Modelling techniques for biological reaction systems 2. Modelling of the steady state case. *Water SA*. 14(4):193–206.

Björnsson, L., Murto, M. & Mattiasson, B. 2000. Evaluation of parameters for monitoring an anaerobic co-digestion process. *Applied Microbiology and Biotechnology*. 54(6):844–849.

Björnsson, L., Murto, M., Jantsch, T.G. & Mattiasson, B. 2001. Evaluation of new methods for the monitoring of alkalinity, dissolved hydrogen and the microbial community in anaerobic digestion. *Water Research*. 35(12):2833–2840.

Boe, K. 2006. Online monitoring and control of the biogas process. Technical University of Denmark.

Boe, K., Batstone, D.J., Steyer, J.P. & Angelidaki, I. 2010. State indicators for monitoring the anaerobic digestion process. *Water Research*. 44(20):5973–5980.

Brouckaert, C.J., Ikumi, D.S. & Ekama, G.A. 2010. A 3 phase anaerobic digestion model. In *12th IWA AD conference*. Guadalajara, Mexico.

Brouckaert, C.J., Brouckaert, B.M. & Ekama, G.A. 2019a. Integration of complete elemental mass balanced stoichiometry and aqueous phase chemistry for bioprocess modelling of liquid and solid waste treatment systems: Part 1 - The physico-chemical framework. *Water SA, Accepted*.

Brouckaert, C.J., Brouckaert, B.M. & Ekama, G.A. 2019b. Integration of complete elemental mass balanced stoichiometry and aqueous phase chemistry for bioprocess modelling of liquid and solid waste treatment systems: Part 5 - Ionic Speciation. *Water SA, Accepted*.

Brown, P.N., Byrne, G.D. & Hindmarsh, A.C. 1989. VODE: A Variable-coefficient ODE Solver. *SIAM J. Sci. Stat. Comput.* 10(5):1038–1051.

Brun, R., Martin, K., Siegrist, H., Gujer, W. & Reichert, P. 2002. Practical identifiability of ASM2d parameters - systematic selection ad tuning of parameter subsets. *Water Research*. 36(16):4113–4127.

Bryers, J.D. 1985. Structured modeling of the anaerobic digestion of biomass particulates. *Biotechnology and Bioengineering*. 27(5):638–649.

Campolongo, F., Cariboni, J. & Saltelli, A. 2007. An effective screening design for sensitivity analysis of large models. *Environmental Modelling and Software*. 22(10):1509–1518.

Chakrabarti, A. & Ghosh, J.K. 2011. *AIC, BIC and Recent Advances in Model Selection*. V. 7. Elsevier B.V. DOI: 10.1016/B978-0-444-51862-0.50018-6.

Chen, Y., Cheng, J.J. & Creamer, K.S. 2008. Inhibition of anaerobic digestion process : A review. *Bioresource Technology*. 99(10):4044–4064.

Chynoweth, D.P., Owens, J.M. & Legrand, R. 2001. Renewable methane from anaerobic digestion of biomass. *Renewable Energy*. 22(1–3):1–8.

Contois, D.E. 1959. Kinetics of bacterial growth: relationship between population density and specific growth rate of continuous cultures. *Journal of General Microbiology*. 21:40–50.

Corominas, L., Rieger, L., Takacs, I., Ekama, G.A., Hauduc, H., Vanrolleghem, P.A., Oehem, A., Gernaey, K.V., et al. 2010. New framework for standardized notation in wastewater treatment modelling. *Water Science and Technology*. 61(4):15.

Cosenza, A., Mannina, G., Vanrolleghem, P.A. & Neumann, M.B. 2013. Global sensitivity analysis in wastewater applications: A comprehensive comparison of different methods. *Environmental Modelling and Software*. 49:40–52.

Costello, D.J., Greenfield, P.F. & Lee, P.L. 1991. Dynamic modelling of a single-stage high-rate anaerobic reactor—II. Model verification. *Water Research*. 25(7):859–871. DOI: 10.1016/0043-1354(91)90167-O.

Dold, P.L., Ekama, G.A. & Marais, G. v. R. 1980. A General Model for the Activated Sludge Process. *Progress in Water Technology*. 12(6):47–72.

Dutta, A., Davies, C. & Ikumi, D.S. 2018. Performance of upflow anaerobic sludge blanket (UASB) reactor and other anaerobic reactor configurations for wastewater treatment: A comparative review and critical updates. *Journal of Water Supply: Research and Technology - AQUA*. 67(8):858–884. DOI: 10.2166/aqua.2018.090.

Ekama, G.A. 2009. Using bioprocess stoichiometry to build a plant-wide mass balance based steady-state WWTP model. *Water Research*. 43(8):2101–2120.

Ekama, G.A. & Brouckaert, C.J. 2019. Integration of complete elemental mass balanced stoichiometry and aqueous phase chemistry for bioprocess modelling of liquid and solid waste treatment systems: Part 4 - Measuring the aqueous phase. *Water SA, Accepted*.

Ekama, G.A., Brouckaert, C.J., Brouckaert, B.M. & Ikumi, D.S. 2019. Integration of complete elemental mass balanced stoichiometry and aqueous phase chemistry for bioprocess modelling of liquid and solid waste treatment systems: Part 2 - Bioprocess stoichiometry. *Water SA, Accepted*.

Fonti, V. 2017. Feature Selection using LASSO. *VU Amsterdam*. 1–26.

Franke-Whittle, I.H., Walter, A., Ebner, C. & Insam, H. 2014. Investigation into the effect of high concentrations of volatile fatty acids in anaerobic digestion on methanogenic communities. *Waste Management*. 34(11):2080–2089.

Friedman, J., Hastie, T. & Tibshirani, R. 2010. Regularization paths for generalized linear models via coordinate descent. *Journal of Statistical Software*. 33(1):1–22.

Giovannini, G., Donoso-Bravo, A., Jeison, D., Chamy, R., Ruíz-Filippi, G. & Vande Wouwer, A. 2016. A review of the role of hydrogen in past and current modelling approaches to anaerobic digestion processes. *International Journal of Hydrogen Energy*. 41(39):17713–17722.

Graef, S. & Andrews, J. 1974. Stability and control of anaerobic digestion. *Water Pollution Control Federation*. 46(4):666–683.

Gujer, W. 2008. *Systems Analysis for Water Technology*. Springer.

Gujer, W. & Zehnder, A.J.B. 1983. Conversion processes in anaerobic digestion. *Water Science and Technology*. 15(8–9):127–167.

Gujer, W., Henze, M., Mino, T., Matsuo, T., Wentzel, M.C. & Marais, G. v. R. 1995. The Activated Sludge Model No. 2: Biological phosphorus removal. *Water Science and Technology*. 31(2):1–11.

Hao, T., Xiang, P., Mackey, H.R., Chi, K., Lu, H., Chui, H., Loosdrecht, M.C.M. Van & Chen, G. 2014. A review of biological sulfate conversions in wastewater treatment. *Water Research*. 65:1–21.

Hao, T., Luo, J., Wei, L., Mackey, H.R., Liu, R., Rey Morito, G. & Chen, G.H. 2015. Physicochemical and biological characterization of long-term operated sulfate reducing granular sludge in the SANI process. *Water Research*. 71:74–84.

Harding, T.H., Ikumi, D.S. & Ekama, G.A. 2011. Incorporating phosphorus into plant wide wastewater treatment plant modelling - Anaerobic digestion. In *8th IWA Watermatex conference*. San Sebastian, Spain, 20 - 22 June.

Hastie, T., Tibshirani, R. & Friedman, J. 2008. *The elements of statistical learning: data mining, inference and prediction*. Springer.

Hauduc, H., Rieger, L., Takács, I., Héduit, A., Vanrolleghem, P. & Gillot, S. 2010. A systematic approach for model verification: application on seven published activated sludge models. *Water Science & Technology*. 61(4):825–39.

Henze, M. & Harremoës, P. 1983. Anaerobic treatment of wastewater in fixed film reactors – A literature review. *Water Science and Technology*. 15(8–9):1–101.

Henze, M., Gujer, W., Mino, T. & van Loosedrecht, M. 2000. *Activated sludge models ASM1, ASM2, ASM2d and ASM3*.

Henze, M., van Loosdrecht, M.C., Ekama, G.A., Brdjanovic, D., Comeau, Y., Amy, G., Garcia, J., Gerba, C., et al. 2008. *Biological wastewater treatment: principles, modelling and design*. M. Henze, Ed. London: London : IWA Pub.

Hey, T., Sandström, D., Ibrahim, V. & Jönsson, K. 2013. Evaluating 5 and 8 pH-point titrations for measuring VFA in full-scale primary sludge hydrolysate. *Water SA*. 39(1):17–22.

Hickey, R., Wu, M., Veiga, M. & Jones, R. 1991. Start-up, operation, monitoring and control of high rate anaerobic treatment systems. *Water Science and Technology*. 24(8):207–255.

Hill, D.T. & Barth, C.L. 1977. A dynamic model for simulation of animal waste digestion. *Journal of the Water Pollution Control Association*. 10(10):2129–2143.

Ikumi, D.S. 2011. The development of a three phase plant-wide mathematical model for sewage treatment. University of Cape Town.

Ikumi, D.S., Brouckaert, C.J. & Ekama, G.A. 2011. A 3 phase anaerobic digestion model. In *8th IWA Watermatex conference*. San Sebastian, Spain. 20 - 22 June.

Ikumi, D.S., Vanrolleghem, P.A., Brouckaert, C.J., Neumann, M.B. & Ekama, G.A. 2013. Towards calibration of phosphorus (P) removal plant-wide models. In *WWTmod2014*. V. 1. 1–9.

Ikumi, D.S., Harding, T.H. & Ekama, G.A. 2014. Biodegradability of wastewater and activated sludge organics in anaerobic digestion. *Water research*. 56:267–79.

Ikumi, D.S., Harding, T.H., Vogts, M., Lakay, M.T., Mafungwa, H.Z., Brouckaert, C.J. & Ekama, G.A. 2015. *Mass balances modelling over wastewater treatment plants III*.

Izzett, H. 1992. The effect of thermophilic heat treatment on the anaerobic digestibility of primary sludge. University of Cape Town.

Jones, R.M. & Takacs, I. 2004. Importance of anaerobic digestion modelling on predicting the overall performance of wastewater treatment plants. In *Proceedings of Anaerobic Digestion*.

Kalyuzhnyi, S., Fedorovich, V., Lens, P., Hulshoff Pol, L. & Lettinga, G. 1998. Mathematical modelling as a tool to study population dynamics between sulfate reducing and methanogenic bacteria. *Biodegradation*. 9(3–4):187–99.

Kalyuzhnyi, S. V., Sklyar, V.I., Davlyatshina, M.A., Parshina, S.N., Simankova, M. V., Kostrikina, N.A. & Nozhevnikova, A.N. 1996. Organic removal and microbiological features of UASB-reactor under various organic loading rates. *Bioresource Technology*. 55(1):47–54.

Kaspar, H.F. & Wuhrmann, K. 1978. Kinetic parameters and relative turnovers of some important catabolic reactions in digesting sludge. *Applied and Environmental Microbiology*. 36(1):1–7.

Kundu, S., Zanganeh, J. & Moghtaderi, B. 2016. A review on understanding explosions from methane-air mixture. *Journal of Loss Prevention in the Process Industries*. 40:507–523.

Lettinga, G., Hulshoff Pol., L.W., Koster, I.W., Wiegnant, W.M., De Zeeuw, W.J., Rinzema, A., Grin, P.C., Roersma, R.E., et al. 1984. High-rate anaerobic waste-water treatment using the UASB Reactor under a wide range of temperature conditions. *Biotechnology and Genetic Engineering Reviews*. 2(9):253–284.

Levenspiel, O. 1972. *Chemical Reaction Engineering*.

Li, D., Chen, L., Liu, X., Mei, Z., Ren, H., Cao, Q. & Yan, Z. 2017. Instability mechanisms and early warning indicators for mesophilic anaerobic digestion of vegetable waste. *Bioresource Technology*. 245(13):90–97.

Li, L., He, Q., Wei, Y., He, Q. & Peng, X. 2014. Early warning indicators for monitoring the process failure of anaerobic digestion system of food waste. *Bioresource Technology*. 171:491–494.

Loewenthal, R.E., Ekama, G.A. & Marais, G. v. R. 1989. Mixed weak acid/base systems Part I - Mixture characterisation. *Water SA*. 15(1):3–24.

Loewenthal, R.E., Wentzel, M.C., Ekama, G.A. & Marais, G. v. R. 1991. Mixed weak acid/base systems Part II: Dosing estimation, aqueous phase. *Water SA*. 17(2):107–122.

Lu, H., Wang, J., Li, S. & Chen, G. 2009. Steady-state model-based evaluation of sulfate reduction, autotrophic denitrification and nitrification integrated (SANI) process. *Water research*. 43(14):3613–3621.

Lu, H., Ekama, G., Wu, D., Feng, J., van Loosdrecht, M.C.M. & Chen, G.-H. 2012a. SANI® process realizes sustainable saline sewage treatment: steady state model-based evaluation of the pilot-scale trial of the process. *Water research*. 46(2):475–90.

Lu, H., Wu, D., Jiang, F., Ekama, G., van Loosdrecht, M.C.M. & Chen, G.H. 2012b. The demonstration of a novel sulfur cycle-based wastewater treatment process: sulfate reduction, autotrophic denitrification, and nitrification integrated (SANI®) biological nitrogen removal process. *Biotechnology and bioengineering*. 109(11):2778–89. DOI: 10.1002/bit.24540.

Lyberatos, G. & Skiadas, I.V. 1999. Modelling of anaerobic digestion - a review. *Global Nest: The International Journal*. 1(2):63–76.

Masse, D.I. & Droste, R.L. 2000. Comprehensive model of anaerobic digestion of swine manure slurry in a sequencing batch reactor. *Water Research*. 34(12):3087–3106.

McCarty, P.L. 1974. Anaerobic processes. *Birmingham Short Course on Design Aspects of Biological Treatment, International Association of Water Pollution Research, Birmingham, England.*

McCarty, P.L. 1975. Stoichiometry of biological reactions. *Progress in Water Technology*. 7(1):157–172.

Moosbrugger, R., Wentzel, M., Ekama, G. & Marais, G. 1993a. Mixed weak acid base chemistry and pH in anaerobic digestion - A review. *Water SA*. 19(1):1–10.

Moosbrugger, R.E., Wentzel, M.C., Ekama, G.A. & Marais, G. v. R. 1993b. Grape wine distillery waste in UASB systems - Feasibility , alkalinity requirements and pH control. *Water SA*. 19(1):53–68.

Moosbrugger, R.E., Wentzel, M.C., Ekama, G.A. & Marais, G. v. R. 1993c. Alkalinity measurement: Part 1 - A 4 pH point titration method to determine the carbonate weak acid/base in an aqueous carbonate solution. *Water SA*. 19(1):11–22.

Moosbrugger, R.E., Wentzel, M.C., Ekama, G.A. & Marais, G. v. R. 1993d. A 5 pH point titration method for determining the carbonate and SCFA weak acid/bases in anaerobic systems. *Water Science and Technology*. 28(2):237–245.

Moosbrugger, R.E., Wentzel, M.C., Ekama, G.A. & Marais, G. v. R. 1993e. Lauter tun (brewery) waste in UASB systems - Feasibility, alkalinity requirements and pH control. *Water SA*. 19(1):41–52.

Morris, M.D. 1991. Factorial sampling plans for preliminary computational experiments. *Technometrics*. 33(2):161–174.

Mosey, F.E. 1983. Mathematical modelling of the anaerobic digestion process: Regulatory mechanisms for the formation of short-chain volatile acids from glucose. *Water Science and Technology*. 15(8–9):209–232.

Musvoto, E.V., Wentzel, M.C., Loewenthal, R.E. & Ekama, G.A. 1997. Kinetic based model for mixed weak acid/base systems. 311–322.

Musvoto, E.V., Ekama, G.A., Wentzel, M.C. & Loewenthal, R.E. 2000. Extension and application of the three-phase weak acid / base kinetic model to the aeration treatment of anaerobic digester liquors. *Water SA*. 26(4):417–438.

Musvoto, E.V., Wentzel, M.C., Loewenthal, R.E. & Ekama, G.A. 2000. Integrated chemical-physical processes modelling - I. Development of a kinetic-based model for mixed weak acid/base systems. *Water Research*. 34(6):1857–1867.

Musvoto, E.V., Wentzel, M.C. & Ekama, G.A. 2000. Integrated chemical-physical process modelling - II. Simulating aeration treatment of anaerobic digester supernatants. *Water Research*. 34(6):1868–1880.

Neumann, M.B. 2012. Comparison of sensitivity analysis methods for pollutant degradation modelling: A case study from drinking water treatment. *Science of the Total Environment*. 433:530–537. DOI: 10.1016/j.scitotenv.2012.06.026.

O'Rourke, J.T. 1968. Kinetics of anaerobic treatment at reduced temperatures. Stanford University.

Parkhurst, D.L. & Appelo, C.A.J. 2013. *Description of input and examples for PHREEQC Version 3 — A computer program for speciation , batch-reaction , one-dimensional transport , and inverse geochemical calculations.*

Pereira, E.L., Milton, C., Campos, M. & Motteran, F. 2013. Physicochemical study of pH, alkalinity and total acidity in a system composed of Anaerobic Baffled Reactor (ABR) in series with Upflow Anaerobic Sludge Blanket reactor (UASB) in the treatment of pig farming wastewater. *Acta Scientiarum Technology*. 35(3):477–483.

Poinapen, J. 2008. Biological sulphate reduction primary sewage sludge as energy source in an upflow anaerobic sludge bed reactor. University of Cape Town.

Poinapen, J. & Ekama, G. 2010a. Biological sulphate reduction with primary sewage sludge in an upflow anaerobic sludge bed reactor – Part 6 : Development of a kinetic model for BSR. *Water SA*. 36(3):203–214.

Poinapen, J. & Ekama, G. 2010b. Biological sulphate reduction with primary sewage sludge in an upflow anaerobic sludge bed reactor – Part 5 : Steady-state model. *Water SA*. 36(3):193–202.

Poinapen, J., Ekama, G. & Wentzel, M. 2009a. Biological sulphate reduction with primary sewage sludge in an upflow anaerobic sludge bed (UASB) reactor – Part 2 : Modification of simple wet chemistry analytical procedures to achieve COD and S mass balances. *Water SA*. 35(5):535–542.

Poinapen, J., Ekama, G. & Wentzel, M. 2009b. Biological sulphate reduction with primary sewage sludge in an upflow anaerobic sludge bed (UASB) reactor – Part 4 :

Bed settling characteristics. *Water SA*. 35(5):553–560.

Poinapen, J., Wentzel, M. & Ekama, G. 2009. Biological sulphate reduction with primary sewage sludge in an upflow anaerobic sludge bed (UASB) reactor – Part 1 : Feasibility study Unites States of America. *Water SA*. 35(5):525–534.

Press, W.H., Flannery, B.P. Saul A. Teukolsky, S.A. & Vetterling, W.T. 2007. *Numerical recipes: the art of scientific computing*. W.H. Press, Ed. Cambridge; Cambridge University Press.

R Core Team. 2019.

Reddi, Y. 2011. Robust multivariable control design: An application to a bank-to-turn missile. University of KwaZulu-Natal.

Ripley, L.E., Boyle, W.C. & Converse, J.C. 1986. Improved alkalimetric monitoring for anaerobic Digestion of high-strength wastes. *Water Pollution Control Federation*. 58(5):406–411.

Ristow, N.E., Sötemann, S.W., Loewenthal, R.E., Wentzel, M.C. & Ekama, G.A. 2005. *Hydrolysis of primary sewage sludge under methanogenic, ccidogenic and sulfate-reducing conditions*.

Romli, M., Keller, J., Lee, P. & Greenfield, P. 1995. Model prediction and verification of a two-stage high-rate anaerobic wastewater treatment system subjected to shock loads. *Process Safety & Environmental Protection*. 73(B2):151–154.

Ruano, M. V., Ribes, J., Ferrer, J. & Sin, G. 2011. Application of the Morris method for screening the influential parameters of fuzzy controllers applied to wastewater treatment plants. *Water Science and Technology*. 63(10):2199–2206.

Saltelli, A., Tarantola, S. & Chan, K.P.. 1999. A quantitative model-independent method for global sensitivity analysis of model output. *Technometrics*. 41(1):39–56.

Sam-Soon, P.A.L.N.S. 1989. Pelletization in the upflow anaerobic sludge bed (UASB) reactor. University of Cape Town.

Sam-Soon, P.A.L.N.S., Loewenthal, R.E., Dold, P.L. & Marais, G. v. R. 1987.

Hypothesis for pelletisation in the upflow anaerobic sludge bed reactor. *Water SA.* 13(2):69–80.

Sam-Soon, P.A.L.N.S., Lowenthal, R.E., Wentzel, M.C. & Marais, G. v. R. 1990. Growth of biopellets on glucose in upflow anaerobic sludge bed(UASB) systems. *Water SA.* 16(3):151–164.

Sam-Soon, P.A.L.N.S., Loewenthal, R.E., Wentzel, M.C. & Marais, G. v. R. 1991a. A long-chain fatty acid, oleate, as sole substrate in Upflow Anaerobic Sludge Bed (UASB) reactor systems. *Water SA.* 17(1):31–36.

Sam-Soon, P.A.L.N.S., Wentzel, M.C., Dold, P.L., Loewenthal, R.E. & Marais, G. v. R. 1991b. Mathematical modelling of upflow anaerobic sludge bed(UASB) systems treating carbohydrate waste waters. *Water SA.* 17(2):91–106.

Seco, A., Ribes, J., Serralta, J. & Ferrer, J. 2004. Biological nutrient removal model No. 1 (BNRM1). *Water Science and Technology Technology.* 50(6):69–78.

Siegrist, H., Vogt, D., Garcia-Heras, J.L. & Gujer, W. 2002. Mathematical model for meso- and thermophilic anaerobic sewage sludge digestion. *Environmental science & technology.* 36(5):1113–23.

Skiadas, I.V., Gavala, H.N. & Lyberatos, G. 2000. Modelling of the periodic anaerobic baffled reactor (PABR) based on the retaining factor concept. *Water Research.* 34(15):3725–3736. DOI: 10.1016/S0043-1354(00)00137-8.

Skoog, D.A., West, D.M., Holler, F.J. & Crouch, S.R. 2004. *Fundamentals of Analytical Chemistry.* (International student edition). Thompson Brooks/Cole.

Sötemann, S.W., Ristow, N.E., Wentzel, M.C. & Ekama, G.A. 2005a. A steady state model for anaerobic digestion of sewage sludges. *Water SA.* 31(4):511–528.

Sötemann, S.W., Van Rensburg, P., Ristow, N.E., Wentzel, M.C., Loewenthal, R.E. & Ekama, G.A. 2005b. Integrated chemical / physical and biological processes modelling Part 2 - Anaerobic digestion of sewage sludges. *Water SA.* 31(4):545–568.

Tait, S., Solon, K., Volcke, E.I.P. & Batstone, D.J. 2012. A unified approach to wastewater chemistry: Model Corrections. In *WWTmod2012.* 1–13.

Thauer, R.K., Jungermann, K. & Decker, K. 1977. Energy conservation in chemotrophic anaerobic bacteria. *Bacteriological reviews.* 41(1):100–180.

Van Rensburg, P., Musvoto, E. V, Wentzel, M.C. & Ekama, G.A. 2003. Modelling multiple mineral precipitation in anaerobic digester liquor. *Water Research.* 37(13):3087–3097.

Vanhooren, H., Meirlaen, J., Amerlinck, Y., Claeys, F., Vangheluwe, H. & Vanrolleghem, P.A. 2003. WEST: Modelling Biological Wastewater Treatment. *Journal of Hydroinformatics.* 05(1):27–50.

Vanrolleghem, P.A., Petersen, B., De Pauw, D.J.W., Dovermann, H., Weijers, S. & Gernaey, K.A. 2003. A comprehensive model calibration procedure for activated sludge models. *Proceedings of the Water Environment Federation, WEFTEC 2003.*

Vavilin, V.A., Rytov, S. V. & Lokshina, L.Y. 1996. A description of hydrolysis kinetics in anaerobic degradation of particulate organic matter. *Bioresource Technology.* 56(2–3):229–237.

Vavilin, V.A., Rytov, S. V, Lokshina, L.Y., Rintala, J.A. & Lyberatos, G. 2001. Simplified hydrolysis models for the optimal design of two-stage anaerobic digestion. *Water research.* 35(17):4247–4251.

Wang, B., Wu, D., Ekama, G.A., Huang, H., Lu, H. & Chen, G.H. 2017. Optimizing mixing mode and intensity to prevent sludge flotation in sulfidogenic anaerobic sludge bed reactors. *Water Research.* 122:481–491.

Wang, J., Lu, H., Chen, G. & Lau, G. 2009. A novel sulfate reduction, autotrophic denitrification, nitrification integrated (SANI) process for saline wastewater treatment. *water research.* 43(9):2363–2372.

Wang, J., Shi, M., Lu, H., Wu, D. & Shao, M. 2011. Microbial community of sulfate-reducing up-flow sludge bed in the SANI® process for saline sewage treatment. *Applied microbiology ....* 90(6):2015–2025.

Water Supplies Department. 2012. *Annual Report 2011/2012.*

Wentzel, M.C., Ekama, G.A., Dold, P.L., Loewenthal, R.E. & Marais, G. v. R. 1988.

Biological Excess Phosporus Removal.

Wentzel, M.C., Moosbrugger, R.E., Sam-Soon, P.A.L.N.S., Ekama, G.A. & Marais, G. v. R. 1994. Tentative guidelines for waste selection, process design, operation and control of upflow anaerobic sludge bed reactors. *Water Science and Technology.* 30(12):31–42.

Wett, B., Takács, I., Batstone, D.J., Wilson, C. & Murthy, S. 2014. Anaerobic model for high-solids or high-temperature digestion - Additional pathway of acetate oxidation. *Water Science & Technology.* 69(8):1634–1640.

van Zyl, P. 2008. Anaerobic digestion of Fischer-Tropsch reaction water: submerged membrane anaerobic reactor design, performance evaluation & modeling. University of Cape Town.

APPENDIX A: DERIVATION OF STOICHIOMETRIC EQUATIONS

1. Stoichiometry in the SANI Anaerobic reactor (SRUSB)

In contrast to the steady-state equations, the WEST® model requires each process to be modelled separately. Thus, the stoichiometric equations below were derived. They are based on (Poinapen & Ekama, 2010a) and (Kalyuzhnyi et al., 1998), but the method of derivation is on a COD/COD basis instead of a mol/mol basis. For each process, there is an associated decay process which decays the biomass. These equations are identical to the methanogenic equations with just the relevant biomass changing. Thus, they are not derived below. Steady-state equations employ the use of the observed yield, E, to account for the growth as well as the endogenous respiration processes, as opposed to the actual biomass yield, Y, used in dynamic models.

Acetogenic sulphidogenesis
Electron Donor Equation

$$\frac{1}{14}CH_3CH_2COO^- + \frac{1}{2}H_2O \rightarrow \frac{3}{14}CO_3{}^{2-} + \frac{5}{14}H^+ + (H^+ + e^-)$$

Anabolic Equation

$$\frac{kY_{ACS}}{Y_B}CO_3{}^{2-} + \frac{nY_{ACS}}{Y_B}NH_4{}^+ + \frac{pY_{ACS}}{Y_B}PO_4{}^{3-} + \frac{(2k + 3p - n)Y_{ACS}}{Y_B}H^+ + Y_{ACS}(H^+ + e^-)$$

$$\rightarrow \frac{Y_{ACS}}{Y_B}C_kH_lO_mN_nP_p + \frac{(3k + 4p - m)Y_{ACS}}{Y_B}H_2O$$

Catabolic Equations

$$\frac{f_{acetate}}{4}CO_3{}^{2-} + \frac{3f_{acetate}}{8}H^+ + f_{acetate}(H^+ + e^-) \rightarrow \frac{f_{acetate}}{8}CH_3COO^- + \frac{f_{acetate}}{2}H_2O$$

$$f_{H_2}(H^+ + e^-) \rightarrow \frac{f_{H_2}}{2}H_2$$

$$\frac{f_{HS^-}}{8}SO_4{}^{2-} + \frac{f_{HS^-}}{8}H^+ + f_{HS^-}(H^+ + e^-) \rightarrow \frac{f_{HS^-}}{8}HS^- + \frac{f_{HS^-}}{2}H_2O$$

where $f_{acetate}, f_{H_2}, f_{HS^-}$ is the fraction of electrons that go to acetate, hydrogen and bisulphide respectively and the three fractions add up to the electrons which are not used to form biomass $(1-Y_{ACS})$. The amount of electrons that go to each of these electron destinations is dictated by literature values from (Poinapen & Ekama, 2010a) and

(Kalyuzhnyi et al., 1998). Thus, $Y_{ACS}, f_{acetate}, f_{H_2}, f_{HS^-}$ add up to 1. Based on this, the overall equation can be formulated as shown below.

Overall Equation

$$CH_3CH_2COO^- + \left(\frac{14kY_{ACS}}{\gamma_B} + \frac{7f_{acetate}}{2} - 3\right)CO_3{}^{2-} + \frac{14nY_{ACS}}{\gamma_B}NH_4{}^+ + \frac{14pY_{ACS}}{\gamma_B}PO_4{}^{3-}$$

$$+ \left(\frac{14Y_{ACS}(2k + 3p - n)}{\gamma_B} + \frac{21f_{acetate}}{4} + \frac{7f_{HS^-}}{4} - 5\right)H^+ + \frac{7f_{HS^-}}{4}SO_4{}^{2-}$$

$$\rightarrow \frac{14Y_{ACS}}{\gamma_B}C_kH_lO_mN_nP_p + \left(\frac{14Y_{ACS}(3k + 4p - m)}{\gamma_B} + 7f_{acetate} + 7f_{HS^-} - 7\right)H_2O$$

$$+ \frac{7f_{acetate}}{4}CH_3COO^- + 7f_{H_2}H_2 + \frac{7f_{HS^-}}{4}HS^-$$

Acetoclastic Sulphidogenesis

Electron Donor Equation

$$\frac{1}{8}CH_3COO^- + \frac{1}{2}H_2O \rightarrow \frac{1}{4}CO_3{}^{2-} + \frac{3}{8}H^+ + (H^+ + e^-)$$

Anabolic Equation

$$\frac{kY_{AS}}{\gamma_B}CO_3{}^{2-} + \frac{nY_{AS}}{\gamma_B}NH_4{}^+ + \frac{pY_{AS}}{\gamma_B}PO_4{}^{3-} + \frac{(2k + 3p - n)Y_{AS}}{\gamma_B}H^+ + Y_{AS}(H^+ + e^-)$$

$$\rightarrow \frac{Y_{AS}}{\gamma_B}C_kH_lO_mN_nP_p + \frac{(3k + 4p - m)Y_{AS}}{\gamma_B}H_2O$$

Catabolic Equation

$$\frac{(1 - Y_{AS})}{8}SO_4{}^{2-} + \frac{(1 - Y_{AS})}{8}H^+ + (1 - Y_{AS})(H^+ + e^-) \rightarrow \frac{(1 - Y_{AS})}{8}HS^- + \frac{(1 - Y_{AS})}{2}H_2O$$

Overall Equation

$$CH_3COO^- + \left(\frac{8kY_{AS}}{\gamma_B} - 2\right)CO_3{}^{2-} + \frac{8nY_{AS}}{\gamma_B}NH_4{}^+ + \frac{8pY_{AS}}{\gamma_B}PO_4{}^{3-} + (1 - Y_{AS})SO_4{}^{2-}$$

$$+ \left(\frac{8Y_{AS}(2k + 3p - n)}{\gamma_B} - Y_{AS} - 2\right)H^+$$

$$\rightarrow (1 - Y_{AS})HS^- + \frac{8Y_{AS}}{\gamma_B}C_kH_lO_mN_nP_p + \left(\frac{8Y_{AS}(3k + 4p - m)}{\gamma_B} - 4Y_{AS}\right)H_2O$$

Hydrogenotrophic Sulphidogenesis

Electron Donor Equation

$$\frac{1}{2}H_2 \rightarrow (H^+ + e^-)$$

Anabolic Equation

$$\frac{kY_{HS}}{\gamma_B}CO_3^{\ 2-} + \frac{nY_{HS}}{\gamma_B}NH_4^{\ +} + \frac{pY_{HS}}{\gamma_B}PO_4^{\ 3-} + \frac{(2k+3p-n)Y_{HS}}{\gamma_B}H^+ + Y_{HS}(H^+ + e^-)$$

$$\rightarrow \frac{Y_{HS}}{\gamma_B}C_kH_lO_mN_nP_p + \frac{(3k+4p-m)Y_{HS}}{\gamma_B}H_2O$$

Catabolic Equation

$$\frac{(1-Y_{HS})}{8}SO_4^{\ 2-} + \frac{(1-Y_{HS})}{8}H^+ + (1-Y_{HS})(H^+ + e^-) \rightarrow \frac{(1-Y_{HS})}{8}HS^- + \frac{(1-Y_{HS})}{2}H_2O$$

Overall Equation

$$H_2 + \frac{2kY_{HS}}{\gamma_B}CO_3^{\ 2-} + \frac{2nY_{HS}}{\gamma_B}NH_4^{\ +} + \frac{2pY_{HS}}{\gamma_B}PO_4^{\ 3-} + \frac{(1-Y_{HS})}{4}SO_4^{\ 2-} + \left(\frac{2Y_{HS}(2k+3p-n)}{\gamma_B}\right.$$

$$\left. + \frac{(1-Y_{HS})}{4}\right)H^+ \rightarrow \frac{2Y_{HS}}{\gamma_B}C_kH_lO_mN_nP_p + \left(\frac{2Y_{HS}(3k+4p-m)}{\gamma_B} + 1 - Y_{HS}\right)H_2O$$

$$+ \frac{(1-Y_{HS})}{4}HS^-$$

2. Stoichiometry in the anoxic MBBR (BAR1) (mediated by the autotrophic denitrifying organisms - ADO)

Electron Donor Equation

$$\frac{1}{8}HS^- + \frac{1}{2}H_2O \rightarrow \frac{1}{8}SO_4^{\ 2-} + \frac{1}{8}H^+ + (H^+ + e^-)$$

Anabolic Equation

$$\frac{kY_{ADO}}{\gamma_B}CO_3^{2-} + \frac{nY_{ADO}}{\gamma_B}NH_4^{+} + \frac{pY_{ADO}}{\gamma_B}PO_4^{3-} + \frac{(2k+3p-n)Y_{ADO}}{\gamma_B}H^{+} + Y_{ADO}(H^{+}+e^{-})$$

$$\rightarrow \frac{Y_{ADO}}{\gamma_B}C_kH_lO_mN_nP_p + \frac{(3k+4p-m)Y_{ADO}}{\gamma_B}H_2O$$

Catabolic Equation

$$\frac{1-Y_{ADO}}{5}NO_3^{-} + \frac{1-Y_{ADO}}{5}H^{+} + (1-Y_{ADO})(H^{+}+e^{-}) \rightarrow \frac{1-Y_{ADO}}{10}N_2 + \frac{3(1-Y_{ADO})}{5}H_2O$$

Overall Equation

$$HS^{-} + \frac{8(1-Y_{ADO})}{5}NO_3^{-} + \frac{8kY_{ADO}}{\gamma_B}CO_3^{2-} + \frac{8nY_{ADO}}{\gamma_B}NH_4^{+} + \frac{8pY_{ADO}}{\gamma_B}PO_4^{3-}$$

$$+ \left(\frac{8Y_{ADO}(2k+3p-n)}{\gamma_B} + \frac{8(1-Y_{ADO})}{5} - 1\right)H^{+} \rightarrow \frac{8Y_{ADO}}{\gamma_B}C_kH_lO_mN_nP_p + SO_4^{2-}$$

$$+ \frac{4(1-Y_{ADO})}{5}N_2 + \left(\frac{24(1-Y_{ADO})}{5} + \frac{8Y_{ADO}(3k+4p-m)}{\gamma_B} - 4\right)H_2O$$

APPENDIX B: DERIVATION OF STEADY STATE EQUATIONS FOR SSSM & SS5M

Typically, steady-state equations are simpler to derive for anaerobic digesters because there is no carry forward of products from a preceding reactor. With the UASB steady-state equations derived hereunder, the inflow of solids causes the equations to be slightly more complex, except for ADR1 which has no inflow of biomass. Consider the nth reactor below. The concentrations entering the nth reactor are the same concentration as that exiting the (n^{th}-1) reactor. Because the ADR1 equations are different from the remainder reactors, these are derived first. The general mass balance formula is as follows (in these equations, for the maximum specific growth rates, the symbol μ instead of mu):

$$\begin{bmatrix} Mass\ change \\ in\ the\ system \end{bmatrix} = \begin{bmatrix} Flux\ of\ compound \\ entering\ the\ system \end{bmatrix} - \begin{bmatrix} Flux\ of\ compound \\ exiting\ the\ system \end{bmatrix}$$

$$+ \begin{bmatrix} Increase\ in\ compound \\ due\ to\ bioprocesses \end{bmatrix} - \begin{bmatrix} Decrease\ in\ compound \\ due\ to\ bioprocesses \end{bmatrix}$$

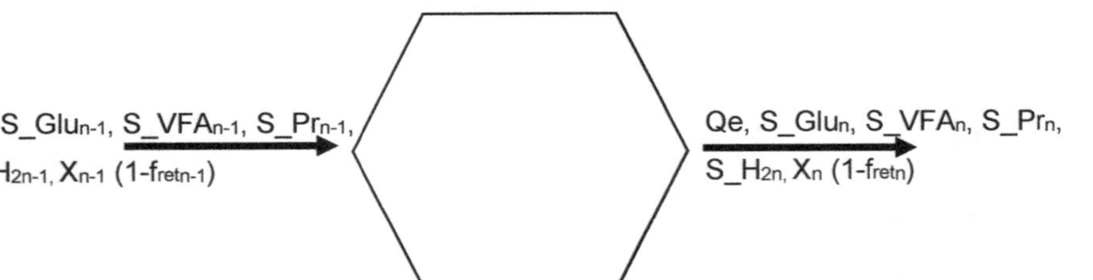

Where $fret_n$ is the solids retention factor in the nth reactor.

Q_i and Q_e are the inflow and outflow respectively

S_VFA, S_Pr, S_H2 and S_Glu are the soluble concentrations of acetate, propionate, hydrogen and glucose respectively (not subject to the effect of $fret_n$)

X_n and X_{n-1} is any particulate concentration such as biomass subject to the effect of $fret_n$ or $fret_{n-1}$

First Anaerobic Digester Reactor Equations (ADR1)

Mass balance on X_{AD}

$$VdX_{AD} = 0 - (1 - f_{ret})Q_e X_{AD}dt + \frac{\mu_{AD}\frac{CS_{Glu}}{MWS_{Glu}}}{K_{S_{AD}} + \frac{CS_{Glu}}{MWS_{Glu}}}CX_{AD}Vdt - b_{AD}CX_{AD}Vdt$$

$\div Vdt$

$$\frac{dX_{AD}}{dt} = -\frac{(1 - f_{ret})Q_e X_{AD}}{V} + \frac{\mu_{AD}\frac{CS_{Glu}}{MWS_{Glu}}}{K_{S_{AD}} + \frac{CS_{Glu}}{MWS_{Glu}}}CX_{AD} - b_{AD}CX_{AD}$$

At steady state, $\frac{dX_{AD}}{dt} = 0$.

$$-\frac{(1-f_{ret})Q_eX_{AD}}{V} + \frac{\mu_{AD}\frac{CS_{Glu}}{MWS_{Glu}}}{K_{S_{AD}} + \frac{CS_{Glu}}{MWS_{Glu}}}CX_{AD} - b_{AD}CX_{AD} = 0$$

$$\frac{\mu_{AD}\frac{CS_{Glu}}{MWS_{Glu}}}{K_{S_{AD}} + \frac{CS_{Glu}}{MWS_{Glu}}}CX_{AD} = \frac{(1-f_{ret})Q_eCX_{AD}}{V} + b_{AD}CX_{AD} \qquad \text{(a)}$$

Mass balance on S_Glu

$$VdS_{Glu} = Q_iCS_{Glu_i}dt - Q_eS_{Glu}dt + \frac{\gamma_oMWS_{Glu}}{\gamma_{Glu}}\frac{CX_{AD}}{MWX_{AD}}(1-f)b_{AD}Vdt - \frac{\mu_{AD}\frac{CS_{Glu}}{MWS_{Glu}}}{K_{S_{AD}} + \frac{CS_{Glu}}{MWS_{Glu}}}\frac{\gamma_oMWS_{Glu}}{Y_{AD}\gamma_{Glu}}\frac{CX_{AD}}{MWX_{AD}}Vdt$$

$$\div Vdt$$

$$\frac{dS_{Glu}}{dt} = \frac{Q_iCS_{Glu_i}}{V} - \frac{Q_eCS_{Glu}}{V} + \frac{\gamma_oMWS_{Glu}}{\gamma_{Glu}}\frac{CX_{AD}}{MWX_{AD}}(1-f)b_{AD} - \frac{\mu_{AD}\frac{CS_{Glu}}{MWS_{Glu}}}{K_{S_{AD}} + \frac{CS_{Glu}}{MWS_{Glu}}}\frac{\gamma_oMWS_{Glu}}{Y_{AD}\gamma_{Glu}}\frac{CX_{AD}}{MWX_{AD}}$$

At steady state, $\frac{dS_{Glu}}{dt} = 0$.

$$\frac{Q_i CS_{Glu_i}}{V} - \frac{Q_e CS_{Glu}}{V} + \frac{\gamma_o MWS_{Glu}}{\gamma_{Glu}} \frac{CX_{AD}}{MWX_{AD}} (1-f) b_{AD} - \frac{\mu_{AD} \frac{CS_{Glu}}{MWS_{Glu}}}{K_{S_{AD}} + \frac{CS_{Glu}}{MWS_{Glu}}} \frac{\gamma_o MWS_{Glu}}{Y_{AD} \gamma_{Glu}} \frac{CX_{AD}}{MWX_{AD}} = 0$$

$$\frac{\mu_{AD} \frac{CS_{Glu}}{MWS_{Glu}}}{K_{S_{AD}} + \frac{CS_{Glu}}{MWS_{Glu}}} CX_{AD} = \left(\frac{Q_i CS_{Glu_i}}{V} - \frac{Q_e CS_{Glu}}{V} + \frac{\gamma_o MWS_{Glu}}{\gamma_{Glu}} \frac{CX_{AD}}{MWX_{AD}} (1-f) b_{AD} \right) \times \frac{MWX_{AD} Y_{AD} \gamma_{Glu}}{\gamma_o MWS_{Glu}}$$

$$\frac{\mu_{AD} \frac{CS_{Glu}}{MWS_{Glu}}}{K_{S_{AD}} + \frac{CS_{Glu}}{MWS_{Glu}}} CX_{AD} = \left(\frac{Q_i CS_{Glu_i}}{V} - \frac{Q_e CS_{Glu}}{V} \right) \times \frac{MWX_{AD} Y_{AD} \gamma_{Glu}}{\gamma_o MWS_{Glu}} + Y_{AD} (1-f) b_{AD} CX_{AD} \qquad \text{(b)}$$

Let (a) = (b) and noting that $Q_e = Q_i$; $\frac{Q_i}{V} = \frac{1}{R_H}$

$$\frac{(1 - f_{ret}) Q_e CX_{AD}}{V} + b_{AD} CX_{AD} = \left(\frac{Q_i CS_{Glu_i}}{V} - \frac{Q_e CS_{Glu}}{V} \right) \times \frac{MWX_{AD} Y_{AD} \gamma_{Glu}}{\gamma_o MWS_{Glu}} + Y_{AD} (1-f) b_{AD} CX_{AD}$$

$$CX_{AD} \left(\frac{(1 - f_{ret})}{R_H} + b_{AD} - Y_{AD} (1-f) b_{AD} \right) = \left(\frac{CS_{Glu_i} - CS_{Glu}}{R_H} \right) \times \frac{MWX_{AD} Y_{AD} \gamma_{Glu}}{\gamma_o MWS_{Glu}}$$

Rearranging gives:

$$CX_{AD} = \left(\frac{CS_{Glu_i} - CS_{Glu}}{(1 - f_{ret}) + b_{AD}R_H(1 - Y_{AD}(1 - f))} \right) \times \frac{MWX_{AD}Y_{AD}Y_{Glu}}{Y_o MWS_{Glu}}$$

From equation (a)

$$\frac{\mu_{AD}\dfrac{CS_{Glu}}{MWS_{Glu}}}{K_{S_{AD}} + \dfrac{CS_{Glu}}{MWS_{Glu}}} = \frac{(1 - f_{ret})Q_e}{V} + b_{AD}$$

$$\mu_{AD}\frac{CS_{Glu}}{MWS_{Glu}} = \left(\frac{(1 - f_{ret})}{R_H} + b_{AD} \right) \times \left(K_{S_{AD}} + \frac{CS_{Glu}}{MWS_{Glu}} \right)$$

$$\mu_{AD}\frac{CS_{Glu}}{MWS_{Glu}} = \frac{(1 - f_{ret})}{R_H}K_{S_{AD}} + b_{AD}K_{S_{AD}} + \frac{(1 - f_{ret})CS_{Glu}}{R_H MWS_{Glu}} + \frac{b_{AD}CS_{Glu}}{MWS_{Glu}}$$

$$\frac{CS_{Glu}}{MWS_{Glu}} \left(\frac{\mu_{AD}R_H - (1 - f_{ret}) - b_{AD}R_H}{R_H} \right) = \frac{(1 - f_{ret})K_{S_{AD}} + b_{AD}K_{S_{AD}}R_H}{R_H}$$

$$CS_{Glu} = \frac{(1 - f_{ret})K_{S_{AD}} + b_{AD}K_{S_{AD}}R_H}{\mu_{AD}R_H - (1 - f_{ret}) - b_{AD}R_H} \times MWS_{Glu}$$

Mass balance on X_{AC}

215

$$VdX_{AC} = 0 - (1 - f_{ret})Q_e X_{AC}dt + \frac{\mu_{AC}\dfrac{CS_{Pr}}{MWS_{Pr}}}{K_{S_{AC}} + \dfrac{CS_{Pr}}{MWS_{Pr}}} \times I_{H2_{AC}} \times CX_{AC}Vdt - b_{AC}CX_{AC}Vdt$$

$\div Vdt$

$$\frac{dX_{AC}}{dt} = -\frac{(1 - f_{ret})Q_e X_{AC}}{V} + \frac{\mu_{AC}\dfrac{CS_{Pr}}{MWS_{Pr}}}{K_{S_{AC}} + \dfrac{CS_{Pr}}{MWS_{Pr}}} \times I_{H2_{AC \times CX_{AC}}} - b_{AC}CX_{AC}$$

At steady state,

$$\frac{dX_{AC}}{dt} = 0$$

$$\frac{\mu_{AC}\dfrac{CS_{Pr}}{MWS_{Pr}}}{K_{S_{AC}} + \dfrac{CS_{Pr}}{MWS_{Pr}}} \times I_{H2_{AC}} \times CX_{AC} = \frac{(1 - f_{ret})Q_e CX_{AC}}{V} + b_{AC}CX_{AC} \qquad \text{(c)}$$

Mass balance on S_Pr

216

$$VdS_{Pr} = 0 - Q_e S_{Pr} dt - MWS_{Pr} \frac{\mu_{AC} \frac{CS_{Pr}}{MWS_{Pr}}}{K_{S_{AC}} + \frac{CS_{Pr}}{MWS_{Pr}}} \frac{\gamma_o \times I_{H2_{AC}}}{MWX_{AC} \times Y_{AC} \times \gamma_{Pr}} CX_{AC} Vdt$$

$$+ MWS_{Pr}(1 - Y_{AH}) \times \frac{\mu_{AD} \frac{CS_{Glu}}{MWS_{Glu}}}{K_{S_{AD}} + \frac{CS_{Glu}}{MWS_{Glu}}} \frac{\gamma_o \times A_{H2_{AD}}}{MWX_{AD} \times Y_{AH} \times \gamma_{Glu}} CX_{AD} Vdt$$

$\div Vdt$

$$\frac{dS_{Pr}}{dt} = -\frac{Q_e S_{Pr}}{V} - MWS_{Pr} \frac{\mu_{AC} \frac{CS_{Pr}}{MWS_{Pr}}}{K_{S_{AC}} + \frac{CS_{Pr}}{MWS_{Pr}}} \frac{\gamma_o \times I_{H2_{AC}}}{MWX_{AC} \times Y_{AC} \times \gamma_{Pr}} CX_{AC} + MWS_{Pr}(1 - Y_{AH}) \times \frac{\mu_{AD} \frac{CS_{Glu}}{MWS_{Glu}}}{K_{S_{AD}} + \frac{CS_{Glu}}{MWS_{Glu}}} \frac{\gamma_o \times A_{H2_{AD}}}{MWX_{AD} \times Y_{AH} \times \gamma_{Glu}} CX_{AD}$$

At steady state, $\frac{dS_{Pr}}{dt} = 0$

$$MWS_{Pr} \frac{\mu_{AC} \frac{CS_{Pr}}{MWS_{Pr}}}{K_{S_{AC}} + \frac{CS_{Pr}}{MWS_{Pr}}} \frac{\gamma_o \times I_{H2_{AC}}}{MWX_{AC} \times Y_{AC} \times \gamma_{Pr}} CX_{AC} = MWS_{Pr}(1 - Y_{AH}) \times \frac{\mu_{AD} \frac{CS_{Glu}}{MWS_{Glu}}}{K_{S_{AD}} + \frac{CS_{Glu}}{MWS_{Glu}}} \frac{\gamma_o \times A_{H2_{AD}}}{MWX_{AD} \times Y_{AH} \times \gamma_{Glu}} CX_{AD} - \frac{Q_e S_{Pr}}{V}$$

$$\frac{\mu_{AC}\frac{CS_{Pr}}{MWS_{Pr}}}{K_{S_{AC}}+\frac{CS_{Pr}}{MWS_{Pr}}}\times I_{H2_{AC}}\times CX_{AC}=\left(MWS_{Pr}(1-Y_{AH})\times\frac{\mu_{AD}\frac{CS_{Glu}}{MWS_{Glu}}}{K_{S_{AD}}+\frac{CS_{Glu}}{MWS_{Glu}}}\frac{\gamma_o\times A_{H2_{AD}}}{MWX_{AD}\times Y_{AH}\times\gamma_{Glu}}CX_{AD}-\frac{Q_eS_{Pr}}{V}\right)\times\frac{MWX_{AC}\times Y_{AC}\times\gamma_{Pr}}{\gamma_o\times MWS_{Pr}}$$

Let (c)=(d) and noting that $Q_e=Q_i$; $\frac{Q_i}{V}=\frac{1}{R_H}$

$$\frac{(1-f_{ret})CX_{AC}}{R_H}+b_{AC}CX_{AC}=\left(MWS_{Pr}(1-Y_{AH})\times\frac{\mu_{AD}\frac{CS_{Glu}}{MWS_{Glu}}}{K_{S_{AD}}+\frac{CS_{Glu}}{MWS_{Glu}}}\frac{\gamma_o\times A_{H2_{AD}}}{MWX_{AD}\times Y_{AH}\times\gamma_{Glu}}CX_{AD}-\frac{S_{Pr}}{R_H}\right)\times\frac{MWX_{AC}\times Y_{AC}\times\gamma_{Pr}}{\gamma_o\times MWS_{Pr}}\qquad\text{(d)}$$

$$CX_{AC}=\left[\left(MWS_{Pr}(1-Y_{AH})\times\frac{\mu_{AD}\frac{CS_{Glu}}{MWS_{Glu}}}{K_{S_{AD}}+\frac{CS_{Glu}}{MWS_{Glu}}}\frac{\gamma_o\times A_{H2_{AD}}}{MWX_{AD}\times Y_{AH}\times\gamma_{Glu}}CX_{AD}-\frac{S_{Pr}}{R_H}\right)\times\frac{MWX_{AC}\times Y_{AC}\times\gamma_{Pr}}{\gamma_o\times MWS_{Pr}}\right]\times\frac{R_H}{(1-f_{ret})+b_{AC}R_H}$$

From Equation (c)

$$\mu_{AC}\frac{CS_{Pr}}{MWS_{Pr}}\times I_{H2_{AC}}=\left(\frac{(1-f_{ret})}{R_H}+b_{AC}\right)\times\left(K_{S_{AC}}+\frac{CS_{Pr}}{MWS_{Pr}}\right)$$

$$\mu_{AC}\frac{CS_{Pr}}{MWS_{Pr}}\times I_{H2_{AC}}-\frac{(1-f_{ret})CS_{Pr}}{R_HMWS_{Pr}}-\frac{b_{AC}CS_{Pr}}{MWS_{Pr}}=\left(\frac{(1-f_{ret})K_{S_{AC}}}{R_H}+b_{AC}K_{S_{AC}}\right)$$

$$CS_{Pr}\left(\frac{R_H\mu_{AC}\times I_{H2_{AC}}-(1-f_{ret})-b_{AC}R_H}{R_H MWS_{Pr}}\right)=\left(\frac{(1-f_{ret})K_{S_{AC}}+b_{AC}K_{S_{AC}}R_H}{R_H}\right)$$

$$CS_{Pr}=\left(\frac{(1-f_{ret})K_{S_{AC}}MWS_{Pr}+b_{AC}K_{S_{AC}}R_H MWS_{Pr}}{R_H\mu_{AC}\times I_{H2_{AC}}-(1-f_{ret})-b_{AC}R_H}\right)$$

Mass Balance on X_{AM}

$$VdX_{AM}=-(1-f_{ret})Q_e CX_{AM}dt+\frac{\mu_{AM}\frac{CS_{VFA}}{MWS_{VFA}}}{K_{S_{AM}}+\frac{CS_{VFA}}{MWS_{VFA}}}\times I_{pH_{AM}}\times CX_{AM}Vdt-b_{AM}CX_{AM}Vdt$$

$\div Vdt$

$$\frac{dX_{AM}}{dt}=-\frac{(1-f_{ret})CX_{AM}}{R_H}+\frac{\mu_{AM}\frac{CS_{VFA}}{MWS_{VFA}}}{K_{S_{AM}}+\frac{CS_{VFA}}{MWS_{VFA}}}\times I_{pH_{AM}}\times CX_{AM}-b_{AM}CX_{AM}$$

At steady state, $\frac{dX_{AM}}{dt}=0$

$$-\frac{(1-f_{ret})CX_{AM}}{R_H}+\frac{\mu_{AM}\frac{CS_{VFA}}{MWS_{VFA}}}{K_{S_{AM}}+\frac{CS_{VFA}}{MWS_{VFA}}}\times I_{pH_{AM}}\times CX_{AM}-b_{AM}CX_{AM}=0$$

$$\frac{\mu_{AM}\dfrac{CS_{VFA}}{MWS_{VFA}}}{K_{S_{AM}}+\dfrac{CS_{VFA}}{MWS_{VFA}}}\times I_{pH_{AM}}\times CX_{AM}=\frac{(1-f_{ret})CX_{AM}}{R_H}+b_{AM}CX_{AM}$$

(e)

Mass Balance on S_VFA

$$VdS_{VFA}=0-Q_eCS_{VFA}dt+MWS_{VFA}(2-2Y_{AD})\frac{\mu_{AD}\dfrac{CS_{Glu}}{MWS_{Glu}}}{K_{S_{AD}}+\dfrac{CS_{Glu}}{MWS_{Glu}}}\frac{\gamma_o\times I_{H2_{AD}}}{MWX_{AD}\times Y_{AD}\times\gamma_{Glu}}CX_{AD}Vdt$$

$$+MWS_{VFA}(1-Y_{AH})\frac{\mu_{AD}\dfrac{CS_{Glu}}{MWS_{Glu}}}{K_{S_{AD}}+\dfrac{CS_{Glu}}{MWS_{Glu}}}\frac{\gamma_o\times A_{H2_{AD}}}{MWX_{AD}\times Y_{AH}\times\gamma_{Glu}}CX_{AD}Vdt$$

$$+MWS_{VFA}(1-Y_{AC})\frac{\mu_{AC}\dfrac{CS_{Pr}}{MWS_{Pr}}}{K_{S_{AC}}+\dfrac{CS_{Pr}}{MWS_{Pr}}}\frac{\gamma_o\times I_{H2_{AC}}}{MWX_{AC}\times Y_{AC}\times\gamma_{Pr}}CX_{AC}Vdt$$

$$-MWS_{VFA}\times\frac{\mu_{AM}\dfrac{CS_{VFA}}{MWS_{VFA}}}{K_{S_{AM}}+\dfrac{CS_{VFA}}{MWS_{VFA}}}\times I_{pH_{AM}}\times\frac{\gamma_o}{MWX_{AM}\times Y_{AM}\times\gamma_{VFA}}CX_{AM}Vdt$$

$$\div Vdt$$

$$\frac{dS_{VFA}}{dt} = -\frac{CS_{VFA}}{R_H} + MWS_{VFA}(2 - 2Y_{AD})\frac{\mu_{AD}\frac{CS_{Glu}}{MWS_{Glu}}}{K_{S_{AD}} + \frac{CS_{Glu}}{MWS_{Glu}}}\frac{\gamma_o \times I_{H2_{AD}}}{MWX_{AD} \times Y_{AD} \times \gamma_{Glu}}CX_{AD}$$

$$+ MWS_{VFA}(1 - Y_{AH})\frac{\mu_{AD}\frac{CS_{Glu}}{MWS_{Glu}}}{K_{S_{AD}} + \frac{CS_{Glu}}{MWS_{Glu}}}\frac{\gamma_o \times A_{H2_{AD}}}{MWX_{AD} \times Y_{AH} \times \gamma_{Glu}}CX_{AD}$$

$$+ MWS_{VFA}(1 - Y_{AC})\frac{\mu_{AC}\frac{CS_{Pr}}{MWS_{Pr}}}{K_{S_{AC}} + \frac{CS_{Pr}}{MWS_{Pr}}}\frac{\gamma_o \times I_{H2_{AC}}}{MWX_{AC} \times Y_{AC} \times \gamma_{Pr}}CX_{AC}$$

$$- MWS_{VFA} \times \frac{\mu_{AM}\frac{CS_{VFA}}{MWS_{VFA}}}{K_{S_{AM}} + \frac{CS_{VFA}}{MWS_{VFA}}} \times I_{pH_{AM}} \times \frac{\gamma_o}{MWX_{AM} \times Y_{AM} \times \gamma_{VFA}}CX_{AM}$$

At steady state, $\frac{dS_{VFA}}{dt} = 0$

$$\frac{\mu_{AM}\frac{CS_{VFA}}{MWS_{VFA}}}{K_{S_{AM}} + \frac{CS_{VFA}}{MWS_{VFA}}} \times I_{pH_{AM}} \times CX_{AM} = \left((2 - 2Y_{AD})\frac{\mu_{AD}\frac{CS_{Glu}}{MWS_{Glu}}}{K_{S_{AD}} + \frac{CS_{Glu}}{MWS_{Glu}}}\frac{\gamma_o \times I_{H2_{AD}}}{MWX_{AD} \times Y_{AD} \times \gamma_{Glu}}CX_{AD} - \frac{CS_{VFA}}{R_H \times MWS_{VFA}} \right.$$

$$+ (1 - Y_{AH})\frac{\mu_{AD}\frac{CS_{Glu}}{MWS_{Glu}}}{K_{S_{AD}} + \frac{CS_{Glu}}{MWS_{Glu}}}\frac{\gamma_o \times A_{H2_{AD}}}{MWX_{AD} \times Y_{AH} \times \gamma_{Glu}}CX_{AD} + (1 - Y_{AC})\frac{\mu_{AC}\frac{CS_{Pr}}{MWS_{Pr}}}{K_{S_{AC}} + \frac{CS_{Pr}}{MWS_{Pr}}}\frac{\gamma_o \times I_{H2_{AC}}}{MWX_{AC} \times Y_{AC} \times \gamma_{Pr}}CX_{AC} \left. \right) \times \frac{MWX_{AM} \times Y_{AM} \times \gamma_{VFA}}{\gamma_o}$$

$$(f)$$

Let (e)=(f)

$$\frac{(1-f_{ret})CX_{AM}}{R_H} + b_{AM}CX_{AM} = \left((2-2Y_{AD})\frac{\mu_{AD}\frac{CS_{Glu}}{MWS_{Glu}}}{K_{S_{AD}}+\frac{CS_{Glu}}{MWS_{Glu}}}\frac{\gamma_o \times I_{H2_{AD}}}{MWX_{AD}\times Y_{AD}\times \gamma_{Glu}}CX_{AD} \right.$$

$$+ (1-Y_{AH})\frac{\mu_{AD}\frac{CS_{Glu}}{MWS_{Glu}}}{K_{S_{AD}}+\frac{CS_{Glu}}{MWS_{Glu}}}\frac{\gamma_o \times A_{H2_{AD}}}{MWX_{AD}\times Y_{AH}\times \gamma_{Glu}}CX_{AD} - \frac{CS_{VFA}}{R_H \times MWS_{VFA}} + (1-Y_{AC})\frac{\mu_{AC}\frac{CS_{Pr}}{MWS_{Pr}}}{K_{S_{AC}}+\frac{CS_{Pr}}{MWS_{Pr}}}\frac{\gamma_o \times I_{H2_{AC}}}{MWX_{AC}\times Y_{AC}\times \gamma_{Pr}}CX_{AC} \left. \right)$$

$$\times \frac{MWX_{AM}\times Y_{AM}\times \gamma_{VFA}}{\gamma_o}$$

$$CX_{AM} = \left((2-2Y_{AD})\frac{\mu_{AD}\frac{CS_{Glu}}{MWS_{Glu}}}{K_{S_{AD}}+\frac{CS_{Glu}}{MWS_{Glu}}}\frac{\gamma_o \times I_{H2_{AD}}}{MWX_{AD}\times Y_{AD}\times \gamma_{Glu}}CX_{AD} + (1-Y_{AH})\frac{\mu_{AD}\frac{CS_{Glu}}{MWS_{Glu}}}{K_{S_{AD}}+\frac{CS_{Glu}}{MWS_{Glu}}}\frac{\gamma_o \times A_{H2_{AD}}}{MWX_{AD}\times Y_{AH}\times \gamma_{Glu}}CX_{AD} \right.$$

$$\left. - \frac{CS_{VFA}}{R_H \times MWS_{VFA}} + (1-Y_{AC})\frac{\mu_{AC}\frac{CS_{Pr}}{MWS_{Pr}}}{K_{S_{AC}}+\frac{CS_{Pr}}{MWS_{Pr}}}\frac{\gamma_o \times I_{H2_{AC}}}{MWX_{AC}\times Y_{AC}\times \gamma_{Pr}}CX_{AC} \right) \times \frac{MWX_{AM}\times Y_{AM}\times \gamma_{VFA}\times R_H}{\gamma_o \times [(1-f_{ret})+b_{AM}R_H]}$$

From Equation (e)

$$\frac{\mu_{AM}\frac{CS_{VFA}}{MWS_{VFA}}}{K_{S_{AM}}+\frac{CS_{VFA}}{MWS_{VFA}}}\times I_{pH_{AM}}=\frac{(1-f_{ret})}{R_H}+b_{AM}$$

$$\mu_{AM}\frac{CS_{VFA}}{MWS_{VFA}}\times I_{pH_{AM}}=\left(\frac{(1-f_{ret})+b_{AM}R_H}{R_H}\right)\left(K_{S_{AM}}+\frac{CS_{VFA}}{MWS_{VFA}}\right)$$

$$\mu_{AM}\frac{CS_{VFA}}{MWS_{VFA}}\times I_{pH_{AM}}-\left(\frac{(1-f_{ret})+b_{AM}R_H}{R_H\times MWS_{VFA}}\right)CS_{VFA}=\left(\frac{(1-f_{ret})+b_{AM}R_H}{R_H}\right)K_{S_{AM}}$$

$$CS_{VFA}\left(\frac{\mu_{AM}\times I_{pH_{AM}}\times R_H-(1-f_{ret})-b_{AM}R_H}{R_H\times MWS_{VFA}}\right)=\left(\frac{(1-f_{ret})+b_{AM}R_H}{R_H}\right)K_{S_{AM}}$$

$$CS_{VFA}=\left(\frac{(1-f_{ret})K_{S_{AM}}MWS_{VFA}+b_{AM}R_HK_{S_{AM}}MWS_{VFA}}{\mu_{AM}\times I_{pH_{AM}}\times R_H-(1-f_{ret})-b_{AM}R_H}\right)$$

Mass Balance on X$_{HM}$

$$VdX_{HM} = -(1 - f_{ret})Q_e CX_{HM}dt + \frac{\mu_{HM}\dfrac{CS_{H_2}}{MWS_{H_2}}}{K_{S_{HM}} + \dfrac{CS_{H_2}}{MWS_{H_2}}} \times I_{pH_{HM}} \times CX_{HM}Vdt - b_{HM}CX_{HM}Vdt$$

$\div Vdt$

$$\frac{dX_{HM}}{dt} = -\frac{(1 - f_{ret})CX_{HM}}{R_H} + \frac{\mu_{HM}\dfrac{CS_{H2}}{MWS_{H_2}}}{K_{S_{HM}} + \dfrac{CS_{H2}}{MWS_{H_2}}} \times I_{pH_{HM}} \times CX_{HM} - b_{HM}CX_{HM}$$

At steady state, $\frac{dX_{HM}}{dt} = 0$

$$-\frac{(1 - f_{ret})CX_{HM}}{R_H} + \frac{\mu_{HM}\dfrac{CS_{H_2}}{MWS_{H_2}}}{K_{S_{HM}} + \dfrac{CS_{H_2}}{MWS_{H_2}}} \times I_{pH_{HM}} \times CX_{HM} - b_{AM}CX_{AM} = 0$$

$$\frac{\mu_{HM}\dfrac{CS_{H_2}}{MWS_{H_2}}}{K_{S_{HM}} + \dfrac{CS_{H_2}}{MWS_{H_2}}} \times I_{pH_{HM}} \times CX_{HM} = \frac{(1 - f_{ret})CX_{HM}}{R_H} + b_{HM}CX_{HM} \tag{g}$$

Mass Balance on S_H2

$$VdS_{H_2} = 0 - Q_e CS_{H_2} dt + MWS_{H_2}(4 - 4Y_{AD}) \frac{\mu_{AD} \frac{CS_{Glu}}{MWS_{Glu}}}{K_{S_{AD}} + \frac{CS_{Glu}}{MWS_{Glu}}} \frac{\gamma_o \times I_{H2_{AD}}}{MWX_{AD} \times Y_{AD} \times \gamma_{Glu}} CX_{AD} Vdt$$

$$+ MWS_{H_2}(1 - Y_{AH}) \frac{\mu_{AD} \frac{CS_{Glu}}{MWS_{Glu}}}{K_{S_{AD}} + \frac{CS_{Glu}}{MWS_{Glu}}} \frac{\gamma_o \times A_{H2_{AD}}}{MWX_{AD} \times Y_{AH} \times \gamma_{Glu}} CX_{AD} Vdt + MWS_{H_2}(3 - 3Y_{AC}) \frac{\mu_{AC} \frac{CS_{Pr}}{MWS_{Pr}}}{K_{S_{AC}} + \frac{CS_{Pr}}{MWS_{Pr}}} \frac{\gamma_o \times I_{H2_{AC}}}{MWX_{AC} \times Y_{AC} \times \gamma_{Pr}} CX_{AC} Vdt$$

$$- MWS_{H_2} \times \frac{\mu_{HM} \frac{CS_{H_2}}{MWS_{H_2}}}{K_{S_{HM}} + \frac{CS_{H_2}}{MWS_{H_2}}} \times I_{pH_{HM}} \times \frac{\gamma_o}{MWX_{HM} \times Y_{HM} \times \gamma_{H2}} CX_{HM} Vdt$$

$$\div Vdt$$

$$\frac{dS_{H_2}}{dt} = -\frac{CS_{H_2}}{R_H} + MWS_{H_2}(4 - 4Y_{AD}) \frac{\mu_{AD} \frac{CS_{Glu}}{MWS_{Glu}}}{K_{S_{AD}} + \frac{CS_{Glu}}{MWS_{Glu}}} \frac{\gamma_o \times I_{H2_{AD}}}{MWX_{AD} \times Y_{AD} \times \gamma_{Glu}} CX_{AD}$$

$$+ MWS_{H_2}(1 - Y_{AH}) \frac{\mu_{AD} \frac{CS_{Glu}}{MWS_{Glu}}}{K_{S_{AD}} + \frac{CS_{Glu}}{MWS_{Glu}}} \frac{\gamma_o \times A_{H2_{AD}}}{MWX_{AD} \times Y_{AH} \times \gamma_{Glu}} CX_{AD} + MWS_{H_2}(3 - 3Y_{AC}) \frac{\mu_{AC} \frac{CS_{Pr}}{MWS_{Pr}}}{K_{S_{AC}} + \frac{CS_{Pr}}{MWS_{Pr}}} \frac{\gamma_o \times I_{H2_{AC}}}{MWX_{AC} \times Y_{AC} \times \gamma_{Pr}} CX_{AC}$$

$$- MWS_{H_2} \times \frac{\mu_{HM} \frac{CS_{H_2}}{MWS_{H_2}}}{K_{S_{HM}} + \frac{CS_{H_2}}{MWS_{H_2}}} \times I_{pH_{HM}} \times \frac{\gamma_o}{MWX_{HM} \times Y_{HM} \times \gamma_{H2}} CX_{HM}$$

At steady state, $\frac{dS_{H_2}}{dt} = 0$

$$MWS_{H_2} \times \frac{\mu_{HM} \dfrac{CS_{H_2}}{MWS_{H_2}}}{K_{S_{HM}} + \dfrac{CS_{H_2}}{MWS_{H_2}}} \times I_{pH_{HM}} \times \frac{\gamma_o}{MWX_{HM} \times Y_{HM} \times \gamma_{H2}} CX_{HM} = MWS_{H_2}(4 - 4Y_{AD}) \frac{\mu_{AD} \dfrac{CS_{Glu}}{MWS_{Glu}}}{K_{S_{AD}} + \dfrac{CS_{Glu}}{MWS_{Glu}}} \frac{\gamma_o \times I_{H2_{AD}}}{MWX_{AD} \times Y_{AD} \times \gamma_{Glu}} CX_{AD} - \frac{CS_{H_2}}{R_H}$$

$$+ MWS_{H_2}(1 - Y_{AH}) \frac{\mu_{AD} \dfrac{CS_{Glu}}{MWS_{Glu}}}{K_{S_{AD}} + \dfrac{CS_{Glu}}{MWS_{Glu}}} \frac{\gamma_o \times A_{H2_{AD}}}{MWX_{AD} \times Y_{AH} \times \gamma_{Glu}} CX_{AD} + MWS_{H_2}(3 - 3Y_{AC}) \frac{\mu_{AC} \dfrac{CS_{Pr}}{MWS_{Pr}}}{K_{S_{AC}} + \dfrac{CS_{Pr}}{MWS_{Pr}}} \frac{\gamma_o \times I_{H2_{AC}}}{MWX_{AC} \times Y_{AC} \times \gamma_{Pr}} CX_{AC}$$

$$\frac{\mu_{HM} \dfrac{CS_{H_2}}{MWS_{H_2}}}{K_{S_{HM}} + \dfrac{CS_{H_2}}{MWS_{H_2}}} \times I_{pH_{HM}} \times CX_{HM} = \left(-\frac{CS_{H_2}}{R_H \times MWS_{H_2}} + (4 - 4Y_{AD}) \frac{\mu_{AD} \dfrac{CS_{Glu}}{MWS_{Glu}}}{K_{S_{AD}} + \dfrac{CS_{Glu}}{MWS_{Glu}}} \frac{\gamma_o \times I_{H2_{AD}}}{MWX_{AD} \times Y_{AD} \times \gamma_{Glu}} CX_{AD} \right.$$

$$\left. + (1 - Y_{AH}) \frac{\mu_{AD} \dfrac{CS_{Glu}}{MWS_{Glu}}}{K_{S_{AD}} + \dfrac{CS_{Glu}}{MWS_{Glu}}} \frac{\gamma_o \times A_{H2_{AD}}}{MWX_{AD} \times Y_{AH} \times \gamma_{Glu}} CX_{AD} + (3 - 3Y_{AC}) \frac{\mu_{AC} \dfrac{CS_{Pr}}{MWS_{Pr}}}{K_{S_{AC}} + \dfrac{CS_{Pr}}{MWS_{Pr}}} \frac{\gamma_o \times I_{H2_{AC}}}{MWX_{AC} \times Y_{AC} \times \gamma_{Pr}} CX_{AC} \right) \times \frac{MWX_{HM} \times Y_{HM} \times \gamma_{H2}}{\gamma_o} \quad \text{(h)}$$

Let (g) = (h)

$$\frac{(1 - f_{ret})CX_{HM}}{R_H} + b_{HM}CX_{HM} = \left(-\frac{CS_{H_2}}{R_H \times MWS_{H_2}} + (4 - 4Y_{AD})\frac{\mu_{AD}\frac{CS_{Glu}}{MWS_{Glu}}}{K_{S_{AD}} + \frac{CS_{Glu}}{MWS_{Glu}}}\frac{\gamma_o \times I_{H2_{AD}}}{MWX_{AD} \times Y_{AD} \times \gamma_{Glu}}CX_{AD} \right.$$

$$\left. + (1 - Y_{AH})\frac{\mu_{AD}\frac{CS_{Glu}}{MWS_{Glu}}}{K_{S_{AD}} + \frac{CS_{Glu}}{MWS_{Glu}}}\frac{\gamma_o \times A_{H2_{AD}}}{MWX_{AD} \times Y_{AH} \times \gamma_{Glu}}CX_{AD} + (3 - 3Y_{AC})\frac{\mu_{AC}\frac{CS_{Pr}}{MWS_{Pr}}}{K_{S_{AC}} + \frac{CS_{Pr}}{MWS_{Pr}}}\frac{\gamma_o \times I_{H2_{AC}}}{MWX_{AC} \times Y_{AC} \times \gamma_{Pr}}CX_{AC} \right) \times \frac{MWX_{HM} \times Y_{HM} \times \gamma_{H2}}{\gamma_o}$$

$$CX_{HM} = \left((4 - 4Y_{AD})\frac{\mu_{AD}\frac{CS_{Glu}}{MWS_{Glu}}}{K_{S_{AD}} + \frac{CS_{Glu}}{MWS_{Glu}}}\frac{\gamma_o \times I_{H2_{AD}}}{MWX_{AD} \times Y_{AD} \times \gamma_{Glu}}CX_{AD} + (1 - Y_{AH})\frac{\mu_{AD}\frac{CS_{Glu}}{MWS_{Glu}}}{K_{S_{AD}} + \frac{CS_{Glu}}{MWS_{Glu}}}\frac{\gamma_o \times A_{H2_{AD}}}{MWX_{AD} \times Y_{AH} \times \gamma_{Glu}}CX_{AD} \right.$$

$$\left. + (3 - 3Y_{AC})\frac{\mu_{AC}\frac{CS_{Pr}}{MWS_{Pr}}}{K_{S_{AC}} + \frac{CS_{Pr}}{MWS_{Pr}}}\frac{\gamma_o \times I_{H2_{AC}}}{MWX_{AC} \times Y_{AC} \times \gamma_{Pr}}CX_{AC} - \frac{CS_{H_2}}{R_H \times MWS_{H_2}} \right) \times \frac{MWX_{HM} \times Y_{HM} \times \gamma_{H2} \times R_H}{\gamma_o[(1 - f_{ret}) + b_{HM}R_H]}$$

From (g):

$$\frac{\mu_{HM}\frac{CS_{H_2}}{MWS_{H_2}}}{K_{S_{HM}} + \frac{CS_{H_2}}{MWS_{H_2}}} \times I_{pH_{HM}} = \frac{(1 - f_{ret})}{R_H} + b_{HM}$$

227

$$\mu_{HM}\frac{CS_{H_2}}{MWS_{H_2}} \times I_{pH_{HM}} - \left(\frac{(1-f_{ret})}{R_H} + b_{HM}\right)\frac{CS_{H_2}}{MWS_{H_2}} = \left(\frac{(1-f_{ret})}{R_H} + b_{HM}\right)K_{S_{HM}}$$

$$CS_{H_2}\left(\frac{\mu_{HM}I_{pH_{HM}}R_H - (1-f_{ret}) - b_{HM}R_H}{R_H MWS_{H_2}}\right) = \left(\frac{(1-f_{ret}) + b_{HM}R_H}{R_H}\right)K_{S_{HM}}$$

$$CS_{H_2} = \left(\frac{(1-f_{ret}) + b_{HM}R_H}{\mu_{HM}I_{pH_{HM}}R_H - (1-f_{ret}) - b_{HM}R_H}\right)K_{S_{HM}}MWS_{H_2}$$

Second to Fifth Anaerobic Digester Reactors (ADR2 to ADR5)

The second to fifth ADR have the exact same equations with the ADR in question being represented by the nth reactor, and the reactor preceding it being (n^{th} -1). The derivation of the equations follows a similar format to the derivations for the first reactor (ADR1), but due to the carry forward of products from the preceding reactor, the equations are different to ADR1. The most detailed derivation shown is done so for acetate (S_VFA) and the acetoclastic methanogen biomass (X_AM). For the remaining products namely glucose (S_Glu), acidogen biomass (X_AD), propionate (S_Pr), acetogen biomass (X_AC), hydrogen (S_H2) and hydrogenotrophic biomass (X_HM), the equations are just shown, but the method of derivation is the same as the acetate and acetoclastic methanogen biomass derivations. Although the derivation of acetate and acetoclastic methanogen biomass is done first, because they depend on the glucose and propionate concentrations, these concentrations would need to be determined first.

Mass balance on X_{AM_n} (Acetoclastic methanogen biomass in n^{th} reactor)

$$V_n dCX_{AM_n} = Q_i CX_{AM_{n-1}}\left(1 - f_{ret_{n-1}}\right)dt - Q_e CX_{AM_n}\left(1 - f_{ret_n}\right)dt + \frac{\mu_{AM}\dfrac{CS_{VFA_n}}{MWS_{VFA}}}{K_{S_{AM}} + \dfrac{CS_{VFA_n}}{MWS_{VFA}}} \times I_{pH_{AM}} \times CX_{AM_n}V_n dt - b_{AM}CX_{AM_n}V_n dt$$

$\div V_n dt$

$$\frac{dCX_{AM_n}}{dt} = \frac{CX_{AM_{n-1}}\left(1 - f_{ret_{n-1}}\right)}{R_{H_n}} - \frac{CX_{AM_n}\left(1 - f_{ret_n}\right)}{R_{H_n}} + \frac{\mu_{AM}\dfrac{CS_{VFA_n}}{MWS_{VFA}}}{K_{S_{AM}} + \dfrac{CS_{VFA_n}}{MWS_{VFA}}} \times I_{pH_{AM}} \times CX_{AM_n} - b_{AM}CX_{AM_n}$$

At steady state, $\dfrac{dCX_{AM_n}}{dt} = 0$

$$\frac{CX_{AM_{n-1}}\left(1 - f_{ret_{n-1}}\right)}{R_{H_n}} - \frac{CX_{AM_n}\left(1 - f_{ret_n}\right)}{R_{H_n}} + \frac{\mu_{AM}\dfrac{CS_{VFA_n}}{MWS_{VFA}}}{K_{S_{AM}} + \dfrac{CS_{VFA_n}}{MWS_{VFA}}} \times I_{pH_{AM}} \times CX_{AM_n} - b_{AM}CX_{AM_n} = 0$$

$$\frac{\mu_{AM}\dfrac{CS_{VFA_n}}{MWS_{VFA}}}{K_{S_{AM}} + \dfrac{CS_{VFA_n}}{MWS_{VFA}}} \times I_{pH_{AM}} \times CX_{AM_n} = \frac{CX_{AM_n}\left(1 - f_{ret_n}\right)}{R_{H_n}} - \frac{CX_{AM_{n-1}}\left(1 - f_{ret_{n-1}}\right)}{R_{H_n}} + b_{AM}CX_{AM_n} \tag{i}$$

Mass balance on S$_{VFA_n}$ (Acetate in nth reactor)

$$V_n dCS_{VFA_n} = Q_i CS_{VFA_{n-1}} dt - Q_e CS_{VFA_n} dt + MWS_{VFA}(2 - 2Y_{AD}) \frac{\mu_{AD} \frac{CS_{Glu_n}}{MWS_{Glu}}}{K_{S_{AD}} + \frac{CS_{Glu_n}}{MWS_{Glu}}} \frac{\gamma_o \times I_{H2_{AD}}}{MWX_{AD} \times Y_{AD} \times \gamma_{Glu}} CX_{AD_n} V_n dt$$

$$+ MWS_{VFA}(1 - Y_{AH}) \frac{\mu_{AD} \frac{CS_{Glu_n}}{MWS_{Glu}}}{K_{S_{AD}} + \frac{CS_{Glu_n}}{MWS_{Glu}}} \frac{\gamma_o \times A_{H2_{AD}}}{MWX_{AD} \times Y_{AH} \times \gamma_{Glu}} CX_{AD_n} V_n dt + MWS_{VFA}(1 - Y_{AC}) \frac{\mu_{AC} \frac{CS_{Pr_n}}{MWS_{Pr}}}{K_{S_{AC}} + \frac{CS_{Pr_n}}{MWS_{Pr}}} \frac{\gamma_o \times I_{H2_{AC}}}{MWX_{AC} \times Y_{AC} \times \gamma_{Pr}} CX_{AC_n} V_n dt$$

$$- MWS_{VFA} \times \frac{\mu_{AM} \frac{CS_{VFA_n}}{MWS_{VFA}}}{K_{S_{AM}} + \frac{CS_{VFA_n}}{MWS_{VFA}}} \times I_{pH_{AM}} \times \frac{\gamma_o}{MWX_{AM} \times Y_{AM} \times \gamma_{VFA}} CX_{AM_n} V_n dt$$

$$\div V_n dt$$

$$\frac{dS_{VFA}}{dt} = \frac{CS_{VFA_{n-1}}}{R_{H_n}} - \frac{CS_{VFA_n}}{R_{H_n}} + MWS_{VFA}(2 - 2Y_{AD}) \frac{\mu_{AD} \frac{CS_{Glu_n}}{MWS_{Glu}}}{K_{S_{AD}} + \frac{CS_{Glu_n}}{MWS_{Glu}}} \frac{\gamma_o \times I_{H2_{AD}}}{MWX_{AD} \times Y_{AD} \times \gamma_{Glu}} CX_{AD_n}$$

$$+ MWS_{VFA}(1 - Y_{AH}) \frac{\mu_{AD} \frac{CS_{Glu_n}}{MWS_{Glu}}}{K_{S_{AD}} + \frac{CS_{Glu_n}}{MWS_{Glu}}} \frac{\gamma_o \times A_{H2_{AD}}}{MWX_{AD} \times Y_{AH} \times \gamma_{Glu}} CX_{AD_n} + MWS_{VFA}(1 - Y_{AC}) \frac{\mu_{AC} \frac{CS_{Pr_n}}{MWS_{Pr}}}{K_{S_{AC}} + \frac{CS_{Pr_n}}{MWS_{Pr}}} \frac{\gamma_o \times I_{H2_{AC}}}{MWX_{AC} \times Y_{AC} \times \gamma_{Pr}} CX_{AC_n}$$

$$- MWS_{VFA} \times \frac{\mu_{AM} \frac{CS_{VFA_n}}{MWS_{VFA}}}{K_{S_{AM}} + \frac{CS_{VFA_n}}{MWS_{VFA}}} \times I_{pH_{AM}} \times \frac{\gamma_o}{MWX_{AM} \times Y_{AM} \times \gamma_{VFA}} CX_{AM_n}$$

At steady state, $\frac{dCS_{VFA_n}}{dt} = 0$

$$\frac{CS_{VFA_{n-1}}}{R_{H_n}} - \frac{CS_{VFA_n}}{R_{H_n}} + MWS_{VFA}(2 - 2Y_{AD})\frac{\mu_{AD}\frac{CS_{Glu_n}}{MWS_{Glu}}}{K_{S_{AD}} + \frac{CS_{Glu_n}}{MWS_{Glu}}}\frac{\gamma_o \times I_{H2_{AD}}}{MWX_{AD} \times Y_{AD} \times \gamma_{Glu}}CX_{AD_n}$$

$$+ MWS_{VFA}(1 - Y_{AH})\frac{\mu_{AD}\frac{CS_{Glu_n}}{MWS_{Glu}}}{K_{S_{AD}} + \frac{CS_{Glu_n}}{MWS_{Glu}}}\frac{\gamma_o \times A_{H2_{AD}}}{MWX_{AD} \times Y_{AH} \times \gamma_{Glu}}CX_{AD_n}$$

$$+ MWS_{VFA}(1 - Y_{AC})\frac{\mu_{AC}\frac{CS_{Pr_n}}{MWS_{Pr}}}{K_{S_{AC}} + \frac{CS_{Pr_n}}{MWS_{Pr}}}\frac{\gamma_o \times I_{H2_{AC}}}{MWX_{AC} \times Y_{AC} \times \gamma_{Pr}}CX_{AC_n} - MWS_{VFA} \times \frac{\mu_{AM}\frac{CS_{VFA_n}}{MWS_{VFA}}}{K_{S_{AM}} + \frac{CS_{VFA_n}}{MWS_{VFA}}} \times I_{pH_{AM}} \times \frac{\gamma_o}{MWX_{AM} \times Y_{AM} \times \gamma_{VFA}}CX_{AM_n} = 0$$

$$\frac{\mu_{AM}\frac{CS_{VFA_n}}{MWS_{VFA}}}{K_{S_{AM}} + \frac{CS_{VFA_n}}{MWS_{VFA}}} \times I_{pH_{AM}} \times CX_{AM_n} = \left(\frac{CS_{VFA_{n-1}}}{R_{H_n} \times MWS_{VFA}} - \frac{CS_{VFA_n}}{R_{H_n} \times MWS_{VFA}} + (2 - 2Y_{AD})\frac{\mu_{AD}\frac{CS_{Glu_n}}{MWS_{Glu}}}{K_{S_{AD}} + \frac{CS_{Glu_n}}{MWS_{Glu}}}\frac{\gamma_o \times I_{H2_{AD}}}{MWX_{AD} \times Y_{AD} \times \gamma_{Glu}}CX_{AD_n}\right.$$

$$\left. + (1 - Y_{AH})\frac{\mu_{AD}\frac{CS_{Glu_n}}{MWS_{Glu}}}{K_{S_{AD}} + \frac{CS_{Glu_n}}{MWS_{Glu}}}\frac{\gamma_o \times A_{H2_{AD}}}{MWX_{AD} \times Y_{AH} \times \gamma_{Glu}}CX_{AD_n} + (1 - Y_{AC})\frac{\mu_{AC}\frac{CS_{Pr_n}}{MWS_{Pr}}}{K_{S_{AC}} + \frac{CS_{Pr_n}}{MWS_{Pr}}}\frac{\gamma_o \times I_{H2_{AC}}}{MWX_{AC} \times Y_{AC} \times \gamma_{Pr}}CX_{AC_n}\right) \times \frac{MWX_{AM} \times Y_{AM} \times \gamma_{VFA}}{\gamma_o} \quad \text{(ii)}$$

Let (i)=(ii)

$$\frac{CX_{AM_n}\left(1-f_{ret_n}\right)}{R_{H_n}}-\frac{CX_{AM_{n-1}}\left(1-f_{ret_{n-1}}\right)}{R_{H_n}}+b_{AM}CX_{AM_n}=\left(\frac{CS_{VFA_{n-1}}}{R_{H_n}\times MWS_{VFA}}-\frac{CS_{VFA_n}}{R_{H_n}\times MWS_{VFA}}\right.$$

$$+\left(2-2Y_{AD}\right)\frac{\mu_{AD}\dfrac{CS_{Glu_n}}{MWS_{Glu}}}{K_{S_{AD}}+\dfrac{CS_{Glu_n}}{MWS_{Glu}}}\frac{\gamma_o\times I_{H2_{AD}}}{MWX_{AD}\times Y_{AD}\times\gamma_{Glu}}CX_{AD_n}+\left(1-Y_{AH}\right)\frac{\mu_{AD}\dfrac{CS_{Glu_n}}{MWS_{Glu}}}{K_{S_{AD}}+\dfrac{CS_{Glu_n}}{MWS_{Glu}}}\frac{\gamma_o\times A_{H2_{AD}}}{MWX_{AD}\times Y_{AH}\times\gamma_{Glu}}CX_{AD_n}$$

$$+\left(1-Y_{AC}\right)\frac{\mu_{AC}\dfrac{CS_{Pr_n}}{MWS_{Pr}}}{K_{S_{AC}}+\dfrac{CS_{Pr_n}}{MWS_{Pr}}}\frac{\gamma_o\times I_{H2_{AC}}}{MWX_{AC}\times Y_{AC}\times\gamma_{Pr}}CX_{AC_n}\left)\times\frac{MWX_{AM}\times Y_{AM}\times\gamma_{VFA}}{\gamma_o}\right.$$

$$CX_{AM_n}\frac{\left(1-f_{ret_n}\right)+b_{AM}R_{H_n}}{R_{H_n}}=\left[\left(\frac{CS_{VFA_{n-1}}}{R_{H_n}\times MWS_{VFA}}-\frac{CS_{VFA_n}}{R_{H_n}\times MWS_{VFA}}+\left(2-2Y_{AD}\right)\frac{\mu_{AD}\dfrac{CS_{Glu_n}}{MWS_{Glu}}}{K_{S_{AD}}+\dfrac{CS_{Glu_n}}{MWS_{Glu}}}\frac{\gamma_o\times I_{H2_{AD}}}{MWX_{AD}\times Y_{AD}\times\gamma_{Glu}}CX_{AD_n}\right.\right.$$

$$+\left(1-Y_{AH}\right)\frac{\mu_{AD}\dfrac{CS_{Glu_n}}{MWS_{Glu}}}{K_{S_{AD}}+\dfrac{CS_{Glu_n}}{MWS_{Glu}}}\frac{\gamma_o\times A_{H2_{AD}}}{MWX_{AD}\times Y_{AH}\times\gamma_{Glu}}CX_{AD_n}+\left(1-Y_{AC}\right)\frac{\mu_{AC}\dfrac{CS_{Pr_n}}{MWS_{Pr}}}{K_{S_{AC}}+\dfrac{CS_{Pr_n}}{MWS_{Pr}}}\frac{\gamma_o\times I_{H2_{AC}}}{MWX_{AC}\times Y_{AC}\times\gamma_{Pr}}CX_{AC_n}\left)\times\frac{MWX_{AM}\times Y_{AM}\times\gamma_{VFA}}{\gamma_o}\right.$$

$$\left.\left.+\frac{CX_{AM_{n-1}}\left(1-f_{ret_{n-1}}\right)}{R_{H_n}}\right]\right.$$

$$CX_{AM_n} = \left[\left(\frac{CS_{VFA_{n-1}}}{R_{H_n} \times MWS_{VFA}} + (2 - 2Y_{AD}) \frac{\mu_{AD} \frac{CS_{Glu_n}}{MWS_{Glu}}}{K_{S_{AD}} + \frac{CS_{Glu_n}}{MWS_{Glu}}} \frac{\gamma_o \times I_{H2_{AD}}}{MWX_{AD} \times Y_{AD} \times \gamma_{Glu}} CX_{AD_n} - \frac{CS_{VFA_n}}{R_{H_n} \times MWS_{VFA}} \right.$$

$$\left. + (1 - Y_{AH}) \frac{\mu_{AD} \frac{CS_{Glu_n}}{MWS_{Glu}}}{K_{S_{AD}} + \frac{CS_{Glu_n}}{MWS_{Glu}}} \frac{\gamma_o \times A_{H2_{AD}}}{MWX_{AD} \times Y_{AH} \times \gamma_{Glu}} CX_{AD_n} + (1 - Y_{AC}) \frac{\mu_{AC} \frac{CS_{Pr_n}}{MWS_{Pr}}}{K_{S_{AC}} + \frac{CS_{Pr_n}}{MWS_{Pr}}} \frac{\gamma_o \times I_{H2_{AC}}}{MWX_{AC} \times Y_{AC} \times \gamma_{Pr}} CX_{AC_n} \right)$$

$$\left. \times \frac{MWX_{AM} \times Y_{AM} \times \gamma_{VFA}}{\gamma_o} + \frac{CX_{AM_{n-1}}(1 - f_{ret_{n-1}})}{R_{H_n}} \right] \times \frac{R_{H_n}}{(1 - f_{ret_n}) + b_{AM} R_{H_n}}$$

From Equation (i)

$$\mu_{AM} \frac{CS_{VFA_n}}{MWS_{VFA}} \times I_{pH_{AM}} \times CX_{AM_n} = \left(\frac{CX_{AM_n}(1 - f_{ret_n}) - CX_{AM_{n-1}}(1 - f_{ret_{n-1}}) + b_{AM} CX_{AM_n} R_{H_n}}{R_{H_n}} \right) \times \left(K_{S_{AM}} + \frac{CS_{VFA_n}}{MWS_{VFA}} \right)$$

$$\mu_{AM} \frac{CS_{VFA_n}}{MWS_{VFA}} \times I_{pH_{AM}} \times CX_{AM_n} - \left(\frac{CX_{AM_n}(1 - f_{ret_n}) - CX_{AM_{n-1}}(1 - f_{ret_{n-1}}) + b_{AM} CX_{AM_n} R_{H_n}}{R_{H_n}} \right) \times \frac{CS_{VFA_n}}{MWS_{VFA}}$$

$$= \left(\frac{CX_{AM_n}(1 - f_{ret_n}) - CX_{AM_{n-1}}(1 - f_{ret_{n-1}}) + b_{AM} CX_{AM_n} R_{H_n}}{R_{H_n}} \right) \times K_{S_{AM}}$$

$$CS_{VFA_n}\left(\frac{\mu_{AM} \times I_{pH_{AM}} \times CX_{AM_n}}{MWS_{VFA}} - \left(\frac{CX_{AM_n}(1 - f_{ret_n}) - CX_{AM_{n-1}}(1 - f_{ret_{n-1}}) + b_{AM}CX_{AM_n}R_{H_n}}{R_{H_n} \times MWS_{VFA}}\right)\right)$$

$$= \left(\frac{CX_{AM_n}(1 - f_{ret_n}) - CX_{AM_{n-1}}(1 - f_{ret_{n-1}}) + b_{AM}CX_{AM_n}R_{H_n}}{R_{H_n}}\right) \times K_{S_{AM}}$$

$$CS_{VFA_n}\left(\frac{\mu_{AM} \times I_{pH_{AM}} \times CX_{AM_n} \times R_{H_n} - CX_{AM_n}(1 - f_{ret_n}) + CX_{AM_{n-1}}(1 - f_{ret_{n-1}}) - b_{AM}CX_{AM_n}R_{H_n}}{R_{H_n} \times MWS_{VFA}}\right)$$

$$= \left(\frac{CX_{AM_n}(1 - f_{ret_n}) - CX_{AM_{n-1}}(1 - f_{ret_{n-1}}) + b_{AM}CX_{AM_n}R_{H_n}}{R_{H_n}}\right) \times K_{S_{AM}}$$

$$CS_{VFA_n} = \left(\frac{CX_{AM_n}(1 - f_{ret_n}) - CX_{AM_{n-1}}(1 - f_{ret_{n-1}}) + b_{AM}CX_{AM_n}R_{H_n}}{\mu_{AM} \times I_{pH_{AM}} \times CX_{AM_n} \times R_{H_n} - CX_{AM_n}(1 - f_{ret_n}) + CX_{AM_{n-1}}(1 - f_{ret_{n-1}}) - b_{AM}CX_{AM_n}R_{H_n}}\right) \times K_{S_{AM}} \times MWS_{VFA}$$

Mass Balance on X_{AD_n}

$$V_n dCX_{AD_n} = Q_i CX_{AD_{n-1}}(1 - f_{ret_{n-1}})dt - Q_e CX_{AD_n}(1 - f_{ret_n})dt + \frac{\mu_{AD}\dfrac{CS_{Glu_n}}{MWS_{Glu}}}{K_{S_{AD}} + \dfrac{CS_{Glu_n}}{MWS_{Glu}}} \times CX_{AD_n}V_n dt - b_{AD}CX_{AD_n}V_n dt$$

Mass Balance on S_{Glu_n}

$$V dCS_{Glu_n} = Q_i CS_{Glu_{n-1}} dt - Q_e CS_{Glu_n} dt + \frac{\gamma_o MWS_{Glu}}{\gamma_{Glu}} \frac{CX_{AD_n}}{MWX_{AD}} (1-f) b_{AD} V dt - \frac{\mu_{AD} \dfrac{CS_{Glu_n}}{MWS_{Glu}}}{K_{S_{AD}} + \dfrac{CS_{Glu_n}}{MWS_{Glu}}} \frac{\gamma_o MWS_{Glu}}{Y_{AD} \gamma_{Glu}} \frac{CX_{AD_n}}{MWX_{AD}} V dt$$

From which (solving the equations in a similar manner as done for S_{VFA_n} and X_{AM_n} above):

$$CX_{AD_n} = \frac{(1 - f_{ret_{n-1}}) CX_{AD_{n-1}} \gamma_o MWS_{Glu} + MWX_{AD} Y_{AD} \gamma_{Glu} \left(CS_{Glu_{n-1}} - CS_{Glu_n} \right)}{\gamma_o MWS_{Glu} \left[(1 - f_{ret_n}) + b_{AD} R_{H_n} - Y_{AD} (1-f) b_{AD} R_{H_n} \right]}$$

$$CS_{Glu_n} = \frac{(1 - f_{ret_n}) CX_{AD_n} - (1 - f_{ret_{n-1}}) CX_{AD_{n-1}} + b_{AD} R_{H_n} CX_{AD_n}}{\mu_{AD} CX_{AD_n} R_{H_n} MWS_{Glu} \left[(1 - f_{ret_n}) + b_{AD} R_{H_n} - Y_{AD} (1-f) b_{AD} R_{H_n} \right]}$$

Mass Balance on X_{AC_n}

$$V_n dCX_{AC_n} = Q_i CX_{AC_{n-1}} (1 - f_{ret_{n-1}}) dt - Q_e CX_{AC_n} (1 - f_{ret_n}) dt - b_{AC} CX_{AC_n} V_n dt + \frac{\mu_{AC} \dfrac{CS_{Pr_n}}{MWS_{Pr}}}{K_{S_{AC}} + \dfrac{CS_{Pr_n}}{MWS_{Pr}}} \times I_{H2_{AC}} \times CX_{AC_n} V_n dt$$

Mass Balance on S_{Pr_n}

$$V_n dCS_{Pr_n} = Q_i CS_{Pr_{n-1}} dt - Q_e CS_{Pr_n} dt + MWS_{Pr}(1 - Y_{AH}) \frac{\mu_{AD} \dfrac{CS_{Glu_n}}{MWS_{Glu}}}{K_{S_{AD}} + \dfrac{CS_{Glu_n}}{MWS_{Glu}}} \frac{\gamma_o \times A_{H2_{AD}}}{MWX_{AD} \times Y_{AH} \times \gamma_{Glu}} CX_{AD_n} V_n dt$$

$$- MWS_{Pr} \times \frac{\mu_{AC} \dfrac{CS_{Pr_n}}{MWS_{Pr}}}{K_{S_{AC}} + \dfrac{CS_{Pr_n}}{MWS_{Pr}}} \times I_{H2_{AC}} \times \frac{\gamma_o}{MWX_{AC} \times Y_{AC} \times \gamma_{Pr}} CX_{AC_n} V_n dt$$

From which (solving the equations in a similar manner as done for S_{VFA_n} and X_{AM_n} above):

$$CX_{AC_n} = \left[\left(\frac{CS_{Pr_{n-1}}}{R_{H_n} \times MWS_{Pr}} - \frac{CS_{Pr_n}}{R_{H_n} \times MWS_{Pr}} + (1 - Y_{AH}) \frac{\mu_{AD} \dfrac{CS_{Glu_n}}{MWS_{Glu}}}{K_{S_{AD}} + \dfrac{CS_{Glu_n}}{MWS_{Glu}}} \frac{\gamma_o \times A_{H2_{AD}}}{MWX_{AD} \times Y_{AH} \times \gamma_{Glu}} CX_{AD_n} \right) \times \frac{MWX_{AC} \times Y_{AC} \times \gamma_{Pr}}{\gamma_o} \right.$$

$$\left. + \frac{CX_{AC_{n-1}}(1 - f_{ret_{n-1}})}{R_{H_n}} \right] \times \frac{R_{H_n}}{(1 - f_{ret_n}) + b_{AC} R_{H_n}}$$

$$CS_{Pr_n} = \left(\frac{CX_{AC_n}(1 - f_{ret_n}) - CX_{AC_{n-1}}(1 - f_{ret_{n-1}}) + b_{AC} CX_{AC_n} R_{H_n}}{\mu_{AC} \times I_{H2_{AC}} \times CX_{AC_n} \times R_{H_n} - CX_{AC_n}(1 - f_{ret_n}) + CX_{AC_{n-1}}(1 - f_{ret_{n-1}}) - b_{AC} CX_{AC_n} R_{H_n}} \right) \times K_{S_{AC}} \times MWS_{Pr}$$

Mass Balance on X_{HM_n}

$$V_n dCX_{HM_n} = Q_i CX_{HM_{n-1}}\left(1 - f_{ret_{n-1}}\right)dt - Q_e CX_{HM_n}\left(1 - f_{ret_n}\right)dt - b_{HM}CX_{HM_n}V_n dt + \frac{\mu_{HM}\dfrac{CS_{H2_n}}{MWS_{H2}}}{K_{S_{HM}} + \dfrac{CS_{H2_n}}{MWS_{H2}}} \times I_{pH_{HM}} \times CX_{HM_n}V_n dt$$

Mass Balance on S_{H2_n}

$$V_n dCS_{H2_n} = Q_i CS_{H2_{n-1}}dt - Q_e CS_{H2_n}dt + MWS_{H2}(4 - 4Y_{AD})\frac{\mu_{AD}\dfrac{CS_{Glu_n}}{MWS_{Glu}}}{K_{S_{AD}} + \dfrac{CS_{Glu_n}}{MWS_{Glu}}}\frac{\gamma_o \times I_{H2_{AD}}}{MWX_{AD} \times Y_{AD} \times \gamma_{Glu}}CX_{AD_n}V_n dt$$

$$+ MWS_{H2}(1 - Y_{AH})\frac{\mu_{AD}\dfrac{CS_{Glu_n}}{MWS_{Glu}}}{K_{S_{AD}} + \dfrac{CS_{Glu_n}}{MWS_{Glu}}}\frac{\gamma_o \times A_{H2_{AD}}}{MWX_{AD} \times Y_{AH} \times \gamma_{Glu}}CX_{AD_n}V_n dt + MWS_{H2}(3 - 3Y_{AC})\frac{\mu_{AC}\dfrac{CS_{Pr_n}}{MWS_{Pr}}}{K_{S_{AC}} + \dfrac{CS_{Pr_n}}{MWS_{Pr}}}\frac{\gamma_o \times I_{H2_{AC}}}{MWX_{AC} \times Y_{AC} \times \gamma_{Pr}}CX_{AC_n}V_n dt$$

$$- MWS_{H2} \times \frac{\mu_{HM}\dfrac{CS_{H2_n}}{MWS_{H2}}}{K_{S_{HM}} + \dfrac{CS_{H2_n}}{MWS_{H2}}} \times I_{pH_{HM}} \times \frac{\gamma_o}{MWX_{HM} \times Y_{HM} \times \gamma_{H2}}CX_{HM_n}V_n dt$$

From which (solving the equations in a similar manner as done for S_{VFA_n} and X_{AM_n} above):

$$CX_{HM_n} = \left[\left(\frac{CS_{H2_{n-1}}}{R_{H_n} \times MWS_{H2}} + (4 - 4Y_{AD})\frac{\mu_{AD}\frac{CS_{Glu_n}}{MWS_{Glu}}}{K_{S_{AD}} + \frac{CS_{Glu_n}}{MWS_{Glu}}}\frac{\gamma_o \times I_{H2_{AD}}}{MWX_{AD} \times Y_{AD} \times \gamma_{Glu}}CX_{AD_n} - \frac{CS_{H2_n}}{R_{H_n} \times MWS_{H2}}\right.$$

$$+ (1 - Y_{AH})\frac{\mu_{AD}\frac{CS_{Glu_n}}{MWS_{Glu}}}{K_{S_{AD}} + \frac{CS_{Glu_n}}{MWS_{Glu}}}\frac{\gamma_o \times A_{H2_{AD}}}{MWX_{AD} \times Y_{AH} \times \gamma_{Glu}}CX_{AD_n} + (3 - 3Y_{AC})\frac{\mu_{AC}\frac{CS_{Pr_n}}{MWS_{Pr}}}{K_{S_{AC}} + \frac{CS_{Pr_n}}{MWS_{Pr}}}\frac{\gamma_o \times I_{H2_{AC}}}{MWX_{AC} \times Y_{AC} \times \gamma_{Pr}}CX_{AC_n}\right)$$

$$\times \frac{MWX_{HM} \times Y_{HM} \times \gamma_{H2}}{\gamma_o} + \frac{CX_{HM_{n-1}}(1 - f_{ret_{n-1}})}{R_{H_n}}\left.\right] \times \frac{R_{H_n}}{(1 - f_{ret_n}) + b_{HM}R_{H_n}}$$

$$CS_{H2_n} = \left(\frac{\left[CX_{HM_n}(1 - f_{ret_n}) - CX_{HM_{n-1}}(1 - f_{ret_{n-1}}) + b_{HM}CX_{HM_n}R_{H_n}\right] \times K_{S_{HM}} \times MWS_{H2}}{\mu_{HM} \times I_{pH_{HM}} \times CX_{HM_n} \times R_{H_n} - CX_{HM_n}(1 - f_{ret_n}) + CX_{HM_{n-1}}(1 - f_{ret_{n-1}}) - b_{HM}CX_{HM_n}R_{H_n}}\right)$$

From the equations for the second to fifth reactors derived above, it is clear that each equation derived depends on the other associated variable. Glucose concentration in the nth reactor depends on acidogen biomass concentration in the nth reactor and vice versa. This is similarly true for acetate and acetoclastic methanogen biomass concentration, propionate and acetogen biomass concentration and hydrogen and hydrogenotrophic biomass concentration. So, a loop or solver function is required in order to solve the equations. Furthermore, the pathways are defined by the hydrogen concentration which can only be determined after the glucose and propionate concentrations are known. So, an assumed hydrogen concentration is used which allows the inhibition terms to be calculated and, consequently, all the concentrations. From this, the actual hydrogen concentration is found, and there would be an error between the assumed hydrogen concentration and the calculated concentration. Another loop or solver function allows this error to be removed. These equations were programmed into Microsoft Excel spreadsheets and serve as the core for the steady-state spreadsheet models, SSSM and SS5M.

APPENDIX C: DETAILED SENSITIVITY ANALYSIS REPORT

1. SUMMARY

This section contains details of the statistical analyses implemented in order to obtain theoretical and quantitative supports for the calibration of the study variables. The full dataset interrogated 13104 observations for 23 parameters and 19 variables. In statistical jargon, any measurement collected over multiple independent replicates is referred to as a *variable*. Thus, to avoid confusion with the wastewater genre use of the same term, all the 42 measurements in the dataset are subsequently referred to as explanatories. In order to differentiate the combination, parameters are termed *primary* explanatories while variables are cited as *secondary* explanatories. Table C1 contains a quantitative summary of all the explanatories in the dataset and Figure C1 contains a graphical illustration of the pairwise linear relationships between them. The values in the table are supported by the visual illustration in Figure C2. All analyses discussed here were implemented using the R (R Core Team, 2018) statistical freeware and the corresponding code scripts are available from the author upon request.

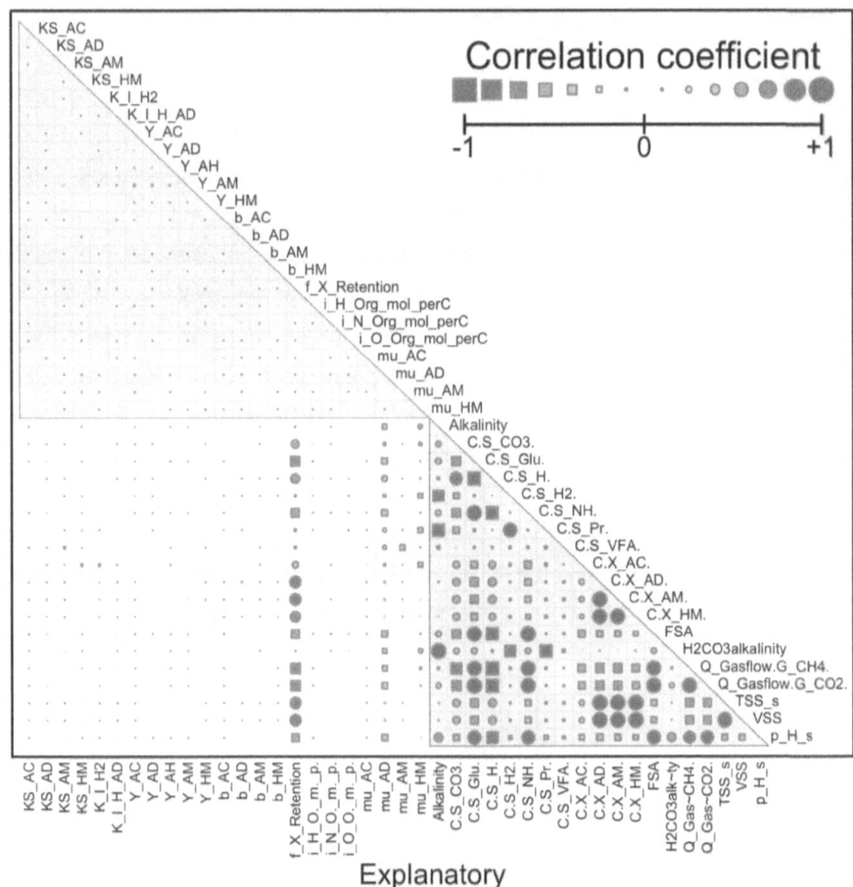

Figure C1: Correlation plot showing the magnitude and the direction of pairwise linear relationships between all the explanatories.

Table C1: Quantitative summary of all the explanatories interrogated in this section. All the estimates are based on the recorded 13104 observations.

Explanatory	Min.	Quantile			Max.	Mean	Std. dev.
		25%	50%	75%			
Primary Explanatory							
KS_AC	0.04	0.07	0.09	0.11	0.13	0.09	0.03
KS_AD	0.39	0.59	0.79	0.97	1.17	0.78	0.22
KS_AM	0.01	0.01	0.01	0.02	0.02	0.01	0.00
KS_HM	0.08	0.12	0.16	0.19	0.23	0.16	0.05
K_I_H2	5.01	7.42	9.94	12.42	15.00	9.94	2.88
K_I_H_AD	0.01	0.01	0.02	0.02	0.02	0.02	0.00
Y_AC	0.04	0.04	0.04	0.04	0.04	0.04	0.00
Y_AD	0.08	0.09	0.09	0.09	0.09	0.09	0.00
Y_AH	0.08	0.09	0.09	0.09	0.09	0.09	0.00
Y_AM	0.04	0.04	0.04	0.04	0.04	0.04	0.00
Y_HM	0.04	0.04	0.04	0.04	0.04	0.04	0.00
b_AC	0.01	0.01	0.02	0.02	0.02	0.02	0.00
b_AD	0.03	0.04	0.04	0.05	0.05	0.04	0.01
b_AM	0.03	0.03	0.04	0.04	0.04	0.04	0.00
b_HM	0.01	0.01	0.01	0.01	0.01	0.01	0.00
f_{ret}	0.50	0.62	0.74	0.86	0.99	0.74	0.14
i_H_Org_mol_perC	1.33	1.36	1.40	1.43	1.47	1.40	0.04
i_N_Org_mol_perC	0.19	0.20	0.20	0.21	0.21	0.20	0.01
i_O_Org_mol_perC	0.38	0.39	0.40	0.41	0.42	0.40	0.01
mu_AC	0.92	1.03	1.15	1.26	1.38	1.15	0.13
mu_AD	0.30	0.47	0.63	0.80	0.96	0.63	0.19
mu_AM	0.88	2.66	4.42	6.10	7.90	4.39	2.01
mu_HM	0.24	0.72	1.21	1.69	2.16	1.21	0.56
Secondary Explanatory							
Alkalinity	0.00	5872.32	5892.56	6058.83	6130.65	5881.12	270.00
C.S_CO3.	0.00	7189.74	7714.20	7732.92	7838.15	7517.46	312.37
C.S_Glu.	0.00	99.32	397.68	4686.62	4691.96	1736.30	2034.43
C.S_H.	0.00	120.03	140.90	141.23	143.86	134.15	9.89
C.S_H2.	0.00	0.01	0.07	0.22	52.09	2.31	9.20
C.S_NH.	0.00	108.85	117.70	173.78	173.87	132.67	28.82
C.S_Pr.	0.00	0.12	1.73	5.61	1688.45	96.19	337.43
C.S_VFA.	-0.01	0.01	0.05	0.11	6.27	0.10	0.20
C.X_AC.	0.00	0.00	14.99	33.84	366.07	25.23	36.83
C.X_AD.	0.00	1.08	939.09	1915.65	11027.43	1430.83	1781.21
C.X_AM.	0.00	0.27	232.50	514.09	3067.23	384.58	500.30
C.X_HM.	0.00	0.00	124.36	292.62	2775.11	244.33	389.28

FSA	0.00	84.52	91.39	134.94	135.01	103.02	22.38
H2CO3alkalinity	0.00	5871.05	5889.80	6042.13	6043.42	5872.96	266.52
Q_Gasflow.G_CH4.	-21.65	-16.07	-14.52	-0.01	0.10	-9.85	7.23
Q_Gasflow.G_CO2.	-50.89	-38.21	-34.61	0.00	0.01	-24.23	17.19
TSS_s	-0.01	2.13	1377.69	2993.67	19978.10	2306.85	3067.92
VSS	-0.01	1.46	1377.38	2993.67	19914.81	2306.59	3067.44
p_H_s	0.00	7.20	7.21	8.19	8.21	7.48	0.47

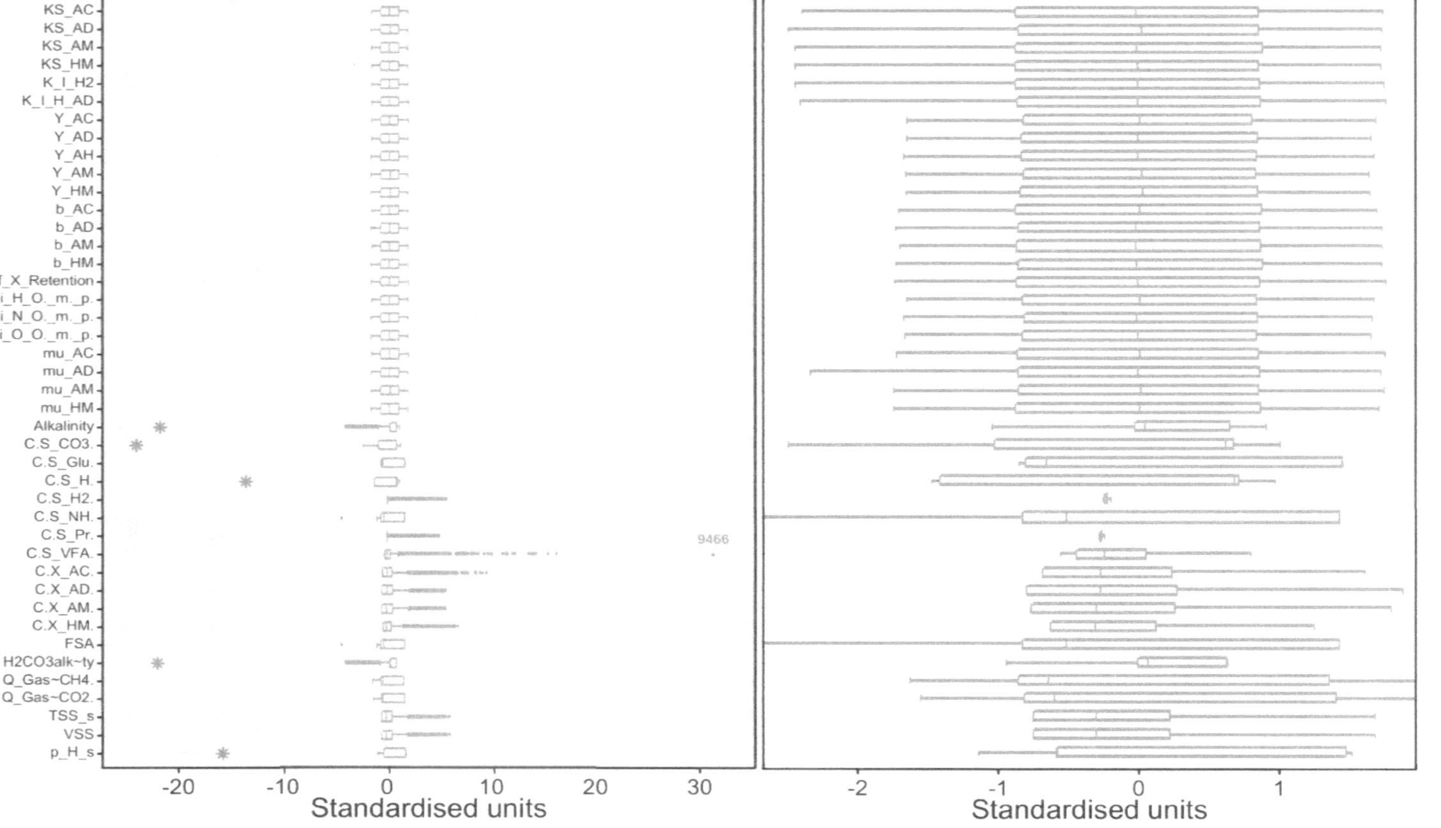

Figure C2: Distributions of the explanatories in the data analysed in this section. The boxplots are generated from standardised values. The illustration on the left includes all the recorded measurements, while the one on the right excludes all the outlying/recur

The minimal linear correlation between the primary explanatory pairs (in the top triangle) in Figure C1 highlights the design of the simulation, where each parameter was randomly assigned for values within the range. The strong correlation evident between the secondary explanatory pairs (in the bottom triangle) is most probably due to the fact that they were mostly products of some expertly generated combination of the primary explanatories. There are some interesting features in the bottom-left rectangle that contains the pairwise relationships between the primary and secondary explanatories. The relationship between f_{ret} and the secondary explanatories is the most pronounced among all its primary counterparts. However, mu_AD, mu_HM and mu_AM also appear to have a slightly significant relationship with the secondary explanatories. Given these apparent relationships between primary and secondary explanatory pairs, f_{ret}, mu_AD, mu_HM and mu_AM are expected to be prominent determinants of the secondary explanatories.

Table C1 contains five-number summaries, means and standard deviations of all the 13 104 records in the full dataset with respect to each of the explanatories considered. Visual support is provided for the table contents in Figure C2, where boxplots illustrating the distributions of the explanatories are presented. Unlike the summaries in the table that were generated with the original data values, the boxes in the figure were constructed with standardised values so that they could all fit on the same graph. It is evident from the table that primary explanatories are all tightly distributed and quite evenly when considering the standardised data. Among the secondary variables, the standard deviation estimates show that C.S_VFA and p_H_s exhibit the least variabilities while TSS_s, VSS and C.S_Glu are characterised with the highest.

There are some interesting features of the data highlighted in Figure C2 that are not immediately evident in Table C1, especially with respect to the secondary explanatories. The shaded regions in the plot on the left-hand in the figure are designed to emphasise "extremely" outlying data points. The use of "extremely" is because the thresholds of the shadings correspond to those of the *median absolute deviations* (MAD) (Leys, Ley, Klein, Bernard, & Licata, 2013) scaled by at least fifty folds [the left shading was drawn up to a hundred fold greater than the normal data cut-off] relative to the recommendations for a normally distributed data. The following deductions are evident from the left plot in the figure.

- Only C.S_VFA has an observation beyond the right threshold. The observation corresponds to the record that has the maximum value under the explanatory. A closer look at Table C1 shows that the said value (6.27 – the record with identifier 9466 in the data) is more than thirty times the standard deviation of C.S_VFA (that is, 6.27/0.20 = 31.35). In fact, if the right threshold were to be extended so that, like its left analogue, it is at least a hundred folds of the prescribed MAD, only C.S_VFA records still appear beyond the band.

- Multiple observations under secondary explanatories Alkalinity, C.S_CO3, C.S_H, H2CO3alkalinity and p_H_s lie beyond the left MAD band. Interestingly, the affected observations – with identifiers 1007, 1524, 2153 and 5623 – are common for all the explanatories, and they are all zeros.

- In order to avoid spurious results, the five observations – that is, records 1007, 1524, 2153, 5623 and 9466 – identified above are deleted from subsequent analyses.

2. METHODOLOGY

The primary aim of this study is to identify the explanatories that have a significant impact on the values of each of the nineteen secondary explanatories. In addition, we sought to quantify the magnitude of the impact. Given the continuity of all the secondary explanatories, ordinary least squares (OLS) regression analyses are adopted as the model of choice. Any introductory statistics textbook such as Rice (2007) will suffice as a good resource on the topic for the interested reader. In terms of the task at hand, the application of OLS is one of explanatory selection and/or regularisation. There are multiple such techniques in the literature and the treatise by Hastie, Tibshirani, & Friedman (2008) provide comprehensive insights about them all.

An OLS regression model is usually expressed as follows.

$$y = \beta_0 + \beta_1 x_1 + \beta_2 x_2 + \beta_3 x_3 + \cdots + \beta_k x_k,$$

Where y is referred to as the response, the x's are referred to as the regressors and the β's are the model coefficients with β_0 taken as the intercept – the average estimate of the response in the absence of the regressors. The model coefficients are estimated by minimising the difference between the left and right hands of the equation. OLS is designed with the aim to formulate the variability observed in the response as a linear combination of the regressors. The principal basis of the model is that the distribution of the response given the regressors is Gaussian. This assumption is traditionally verified by checking that the model residuals are Gaussian distributed. In the context of this project, each of the secondary explanatories is independently considered as a response in a model that includes all the primary and the remainder of the secondary explanatories.

Explanatory selection and regularisation techniques can be classified into two broad groups namely *discrete* and *continuous* methods. The discrete selection method includes (a.) *best subset selection*, where all possible combination of the regressors are fitted against the response and the optimal subset based on the pre-defined quality measure is selected. With respect to this analysis, each of the nineteen secondary explanatory comprises at least forty-one regressors. Thus, there is a need to explore about $2.20 \times 10^{12} \left[= \sum_{i=1}^{41} \binom{41}{i}\right]$ models for each response if the best subset method is used. This is a cumbersome task especially when the complexity increases with the introduction of all the 820 $\left[= \binom{41}{2}\right]$ possible pairwise interaction terms of the regressors into the analysis. Examples of appropriate model quality measure common in the literature include Akaike Information Criterion, Bayes Information Criterion and (Adjusted) R^2. (b.) *forward selection*, where all the OLS one-regressor models are fitted, and the optimal model is chosen based on the appropriate quality measure. All the two-regressor models are then fitted with the chosen explanatory from the one-regressor step kept in the model. An optimal model is again chosen. This inclusion is repeated until a model with all the regressors is fitted. An overall optimal is finally chosen from comparing all the chosen models at each step. (c.) *backward selection*. This technique is similar to forward selection technique except that it begins with the model containing all the regressors before sequentially excluding less significant explanatories. Forward and backward selection techniques are likely to be unable to identify the globally optimal model because they often explore very limited proportions of the model sample space. For example, either approach would only explore 820 $(= \sum_{i=1}^{41} i)$ of the 2.20×10^{12} possible models considered by the best subset technique.

Continuous explanatory regularisation techniques include *partial least squares* (PLS), *principal components regression* (PCR) and *ridge regression* (RR) methods. These methods tend to adjust the regressor coefficients as a function of the magnitude of their relevance in the regression model based on pre-defined penalty factors. PLS and PCR accept integer penalty values within the range of zero and the number of model regressors, while RR accepts fraction within the range $[0, \infty)$. For PCR and PLS, the smaller the penalty values, the more the constraint applied to the coefficients such that when the penalty is equal to the number of model regressors, the values of the coefficients should be equal to those of an analogous OLS model. The behaviour of the penalty factor for RR is vice-versa. That is, for RR, the smaller the penalty, the less constrained the model coefficients, and analogous OLS estimates should be obtained with a penalty of zero. Another distinguishing attribute of these techniques is that PLS and PCR sometimes inflate the coefficients attached to regressors with significant impact on the response beyond their OLS estimates and shrinks those of low impact regressors. Contrarily, RR simply relaxes the degree of coefficient shrinkage until it converges to the analogous OLS estimates. Unlike the discrete selection methods where important explanatories could be inferred as the terms present in the final reduced model, continuous regularisation techniques do not exclude regressors. As a result, explanatory importance inference is less straightforward with continuous regularisation approaches thereby rendering them less appealing to the task at hand.

Some preliminary analyses were carried out to examine the feasibility of applying PCR to achieve the tasks set out in this study. The results are not presented here because, as mentioned above, inference about regressor importance is not straightforward. A noteworthy observation was however made. The analyses were executed with two forms of the data – an original and a standardised version. The use of standardised data follows the norm in wastewater literature; wherein sensitivity analyses are conducted. The non-standardised data always required less penalty value compared to the standardised dataset to achieve an adjusted R^2 value of at least 0.70. This experiment supports the choice in this work not to standardise the data as well a claim that unjustifiably standardising analysis data causes unintended loss of information. Compared to nominal R^2, the adopted adjusted analogue penalises more complex models. The use of 0.70 is arbitrary and is without loss of generality. It implies that 70% of the variability in the response is explained by the corresponding model. This adjusted R^2 lower-bound is maintained throughout this study.

There is also a semi-discrete explanatory regularisation technique known as the *lasso*. This method combines the coefficient shrinkage property of the continuous regularisation approaches and the explanatory exclusion feature of the discrete selection methods. In other words, the method shrinks regressors based on their significance in the model, and it excludes insignificant regressors. The degree of shrinkage and exclusion depends on specified penalty value that could be any fraction between zero and infinity like RR, the smaller the penalty, the less the constraint on the coefficients. Unlike the fully continuous regularisation techniques, regressor importance inference is simplified in lasso due to its exclusion property. Consequently, the lasso is the most suitable approach that is subsequently used here.

3. DATA ANALYSIS

The dataset used here is the one free of the extreme outliers identified in the exploratory analysis section. It contains 13099 observations but retains forty-two explanatories like the original dataset. As explained in the previous section, the majority of the secondary explanatories are direct products of their primary analogues. Therefore, the premier quest is to verify if there exists any linear combination of the primary explanatories that provides "quality" explanation of the variability evident in each of the secondary explanatories. The use of "quality" is to emphasise that the sought model must satisfy certain benchmark criterion. In this case, the chosen threshold is an adjusted R^2 value of 0.70. When there are multiple models that satisfy the adjusted R^2 test, an optimal model is then selected based on three additional criteria. These rules include, model parsimony, expert judgement and/or the Akaike Information Criterion (AIC). Expert judgements were necessary for because, when adjusted R^2 values are high (greater than about 0.85, say) AIC was observed to be often unreliable. Models that are more realistic as judged by the degree of literature support, or models that are simpler or those with lower AICs are preferred.

Prior to deciding on models of choice, three versions of regressor dataset were analysed for each of the secondary explanatory response. This was necessary to ensure that the final model is globally optimal. The first version (D_p) is the data containing only the primary explanatories. The second version $(D_{p\odot})$ is one that contains only the primary explanatories and all the possible pairwise interaction terms between them. The third regressor data (D_F) is a set of all the primary and secondary explanatories, excluding the response of interest. Note that the implementation of lasso with interaction terms here ensures that when a regressor is excluded from a model, all the interaction terms that depend on it are also removed. This is important to ensure that the inferences are interpretable and valid. Given the strong linear correlation observed between the secondary explanatory, the models are guaranteed to yield adjusted R^2 values that exceed the 70% threshold, at least for D_F where secondary explanatories are introduced.

The components of the lasso inferred models are presented in Table C2. The procedure was implemented using the glmnet (Friedman, Hastie, & Tibshirani, 2008) package. For improved accuracy, two sets that each contained 201 evenly spaced penalty values were considered for individual response. The first set contained values in the $[0; 10]$ range, while the second had values in the $[0; 1000]$ range. Simple models were obtained for the majority of the secondary explanatories. The following deductions follow from the contents of Table C2.

The challenge with the results in Table C2 is that while the adjusted R^2 obtained exceeds the prescribed value of 0.7, the causality of the primary explanatories is overshadowed by the association between the secondary explanatories. This is evident from Appendix D, wherein the graphs of the coefficients are plotted. Consequently, the analysis was repeated using only primary explanatories in order to draw inference about their importance without the effect of association. These results are presented in Appendix D.

Table C2: Optimal components for each of the secondary explanatories as a function of the other measurements in the data. The reported penalty values were produced in the corresponding model.

Output Variable	Explanatories	Lasso penalty	Adj R^2
Alkalinity	KS_AM, K_I_H2, Y_AD b_AC, b_AM, b_HM, i_N_Org_mol_perC, i_O_Org_mol_perC, mu_AC, mu_HM, S_CO3, S_Glu, S_H, S_H2, S_NH, S_Pr, X_AC, TSS, p_H	0.1	0.999
S_CO3	Y_AD, b_AC, b_AD, i_H_Org_mol_perC, i_N_Org_mol_perC, mu_AD, Alkalinity, S_Glu, S_H, S_NH, S_Pr, X_AC, X_AM, X_HM, Gasflow_CO2	0.1	1.00
S_H2	K_I_H2, b_AC, f_ret, mu_AC, mu_HM, S_H, S_Pr, X_AC, Q_GasflowG_CH4	0.1	0.858413
S_Pr	KS_AD, b_AC, f_ret, i_N_Org_mol_perC, mu_AD, mu_HM, Alkalinity, S_CO3, S_H, S_H2, S_NH, X_AC, X_HM, Q_GasflowG_CH4, TSS, p_H	0.45	1
S_VFA	KS_AC, KS_AD, KS_AM, KS_HM, K_I_H2, K_I_H_AD, Y_AC, Y_AD, Y_AH, Y_AM, Y_HM, b_AC, b_AD, b_AM, b_HM, f_ret, i_H_Org_mol_perC, i_N_Org_mol_perC, i_O_Org_mol_perC, mu_AC, mu_AD, mu_AM, mu_HM, Alkalinity, S_CO3, S_Glu, S_H, S_H2, S_NH, S_Pr, X_AC, X_AD, X_AM, X_HM, Gasflow_CH4, Gasflow_CO2, TSS, p_H	0.00	0.926163
X_AC	KS_AC, KS_AD, KS_HM, K_I_H2, K_I_H_AD, Y_AC, Y_AD, Y_AH, Y_AM, Y_HM, b_AC, b_AM, b_HM, f_ret, i_H_Org_mol_perC, i_N_Org_mol_perC, i_O_Org_mol_perC, mu_AC, mu_AM, mu_HM, S_Glu, S_H, S_H2, S_NH, S_Pr, S_VFA, X_AD, X_AM, X_HM, TSS	5.00	0.700
X_AD	Y_AD, b_AD, f_ret, Alkalinity, S_Glu, Gasflow_CO2, TSS	9.65	0.997901323
X_AM	Y_AD, Y_AM, b_AM, f_ret, S_CO3, S_Pr, TSS	5.3	0.99468
X_HM	Y_HM, f_ret, Alkalinity, S_NH, TSS	5.00	0.989449
TSS	Y_AD, Y_HM, b_AD, b_AM, i_O_Org_mol_perC, mu_HM, S_H2, S_NH, S_Pr, X_AC, X_AD, X_AM, X_HM	1.9	0.99967

APPENDIX D: FULL SENSITIVITY RESULTS

In this section, the results from the sensitivity analysis are provided. In the case of glucose, pH, methane gas flow and carbon dioxide gas flow where the adjusted R^2 was greater than 0.7, no additional linear model fitting (such as with interactions or secondary explanatories) was done. For the remainder of the output variables, the models were increased in complexity by the inclusion of parameter pairwise interactions, followed by the inclusion of other variables without pairwise interactions. For the more complex models which included secondary explanatories or interaction, the lasso method was used to identify the most important parameters. Because the parameter models without interaction do not include secondary explanatories, which add the complexity of causality vs association, and so are significantly simpler to analyse, the lasso method was not used in these models to identify important explanatories. Lasso was only implemented for the models which include interaction and secondary explanatories. Lasso results, because they contain so few explanatories, are shown in table format with the penalty factors and coefficients. The full model results (before lasso) are shown in graph format (due to the large number of explanatories) as are the Morris Screening results. The analysis results hereunder, therefore include:

1)   Full models using only primary explanatories, no interaction (Graphs)

2)   Lasso results for models using primary explanatories with interaction (Table)

3)   Full models using both primary and secondary explanatories (Graphs)

4)   Lasso results for models with primary and secondary explanatories (Table)

5)   Morris Screening (Graphs)

1. FULL MODELS WITH ONLY PRIMARY EXPLANATORIES, NO INTERACTION

The graphs below are compiled using only primary explanatories and no interaction. Although these models may not all meet the requirement of an adjusted R^2 greater than 0.7, it serves as a comparison between the more complex models. Because the dataset has been standardised, the highest β coefficients point to the explanatory with the largest effect on the overall model.

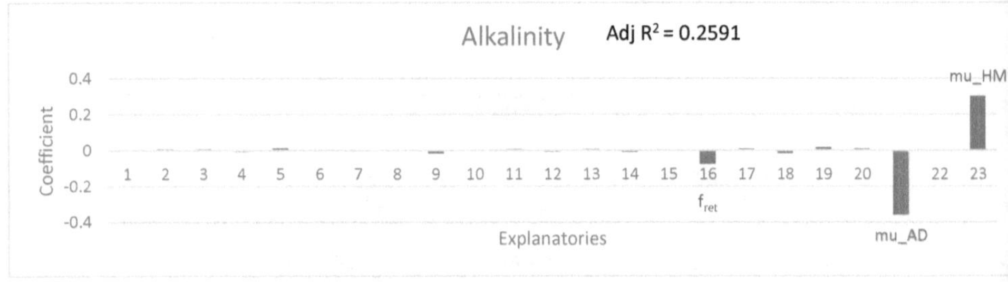

Figure D1: β coefficients for Alkalinity

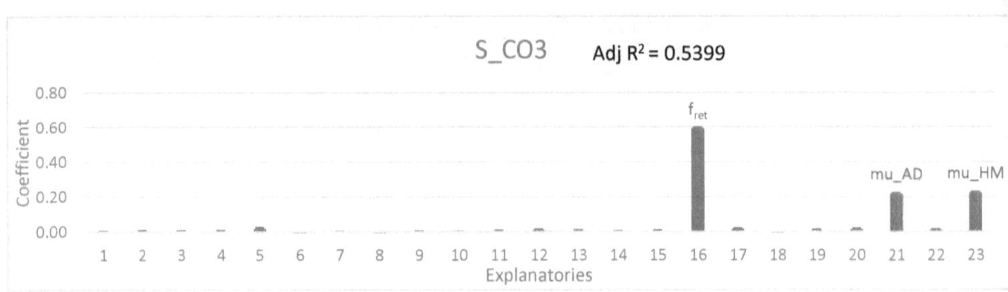

Figure D2: β coefficients for S_CO3

Figure D3: β coefficients for S_Glu

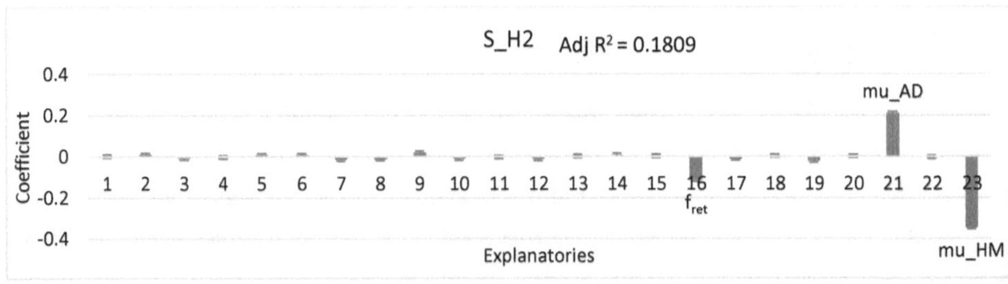

Figure D4: β coefficients for S_H2

Figure D6: β coefficients for S_H

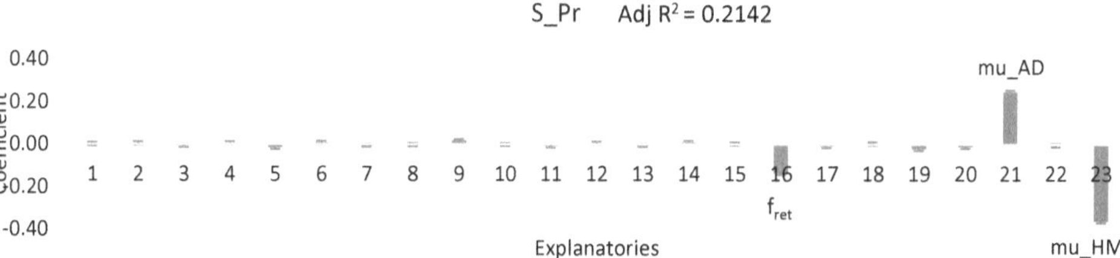

Figure D5: β coefficients for S_Pr

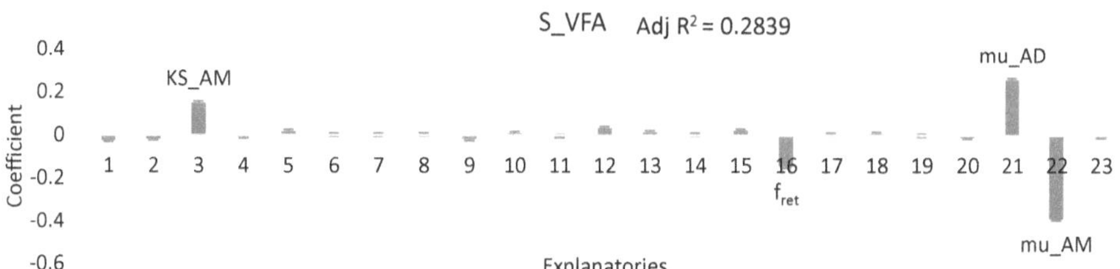

Figure D7: β coefficients for S_VFA

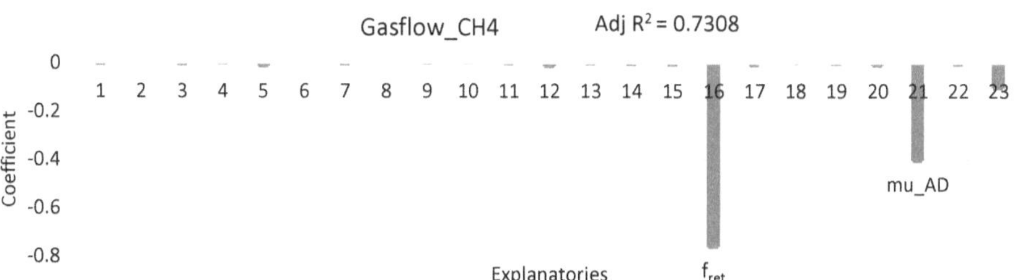

Figure D8: β coefficients for Gasflow_CH4

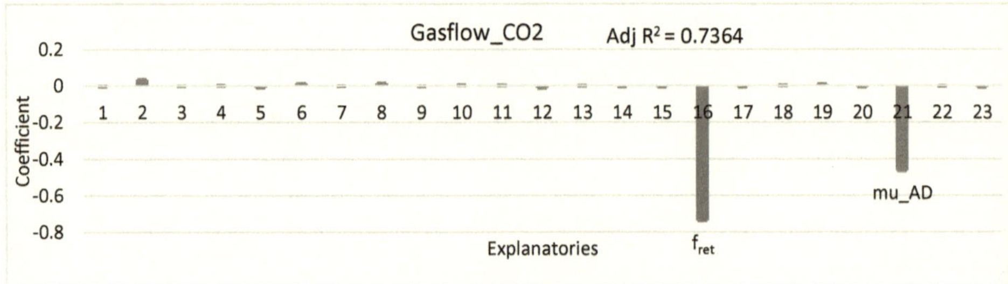

Figure D9: β coefficients for Gasflow_CO2

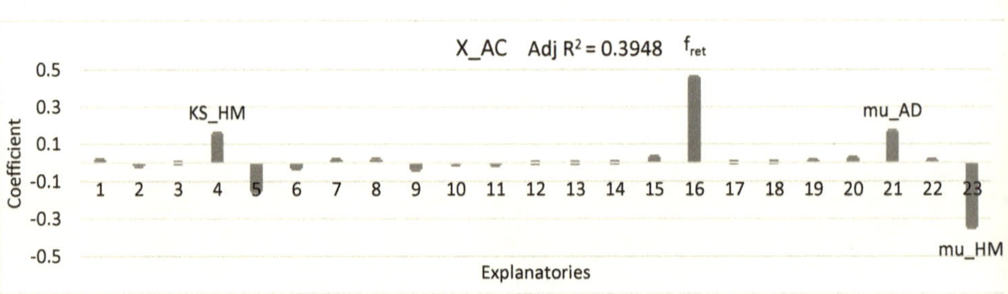

Figure D10: β coefficients for X_AC

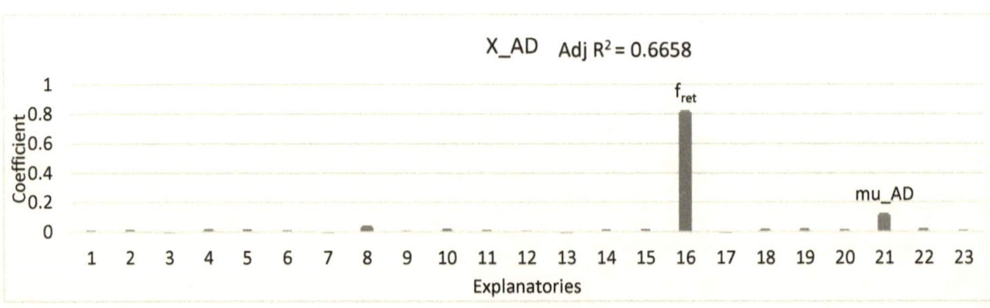

Figure D11: β coefficients for X_AD

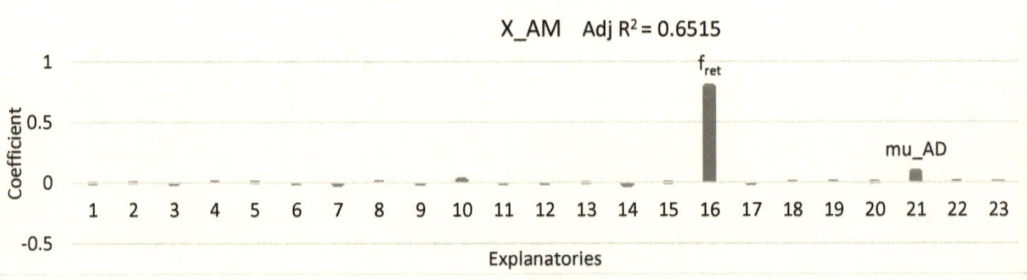

Figure D12: β coefficients for X_AM

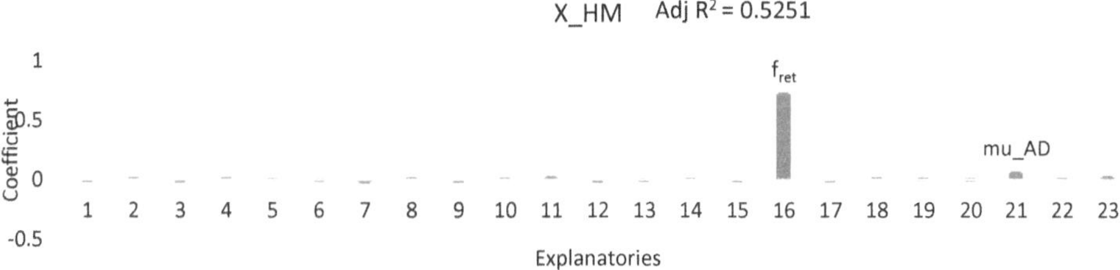

Figure D13: β coefficients for X_HM

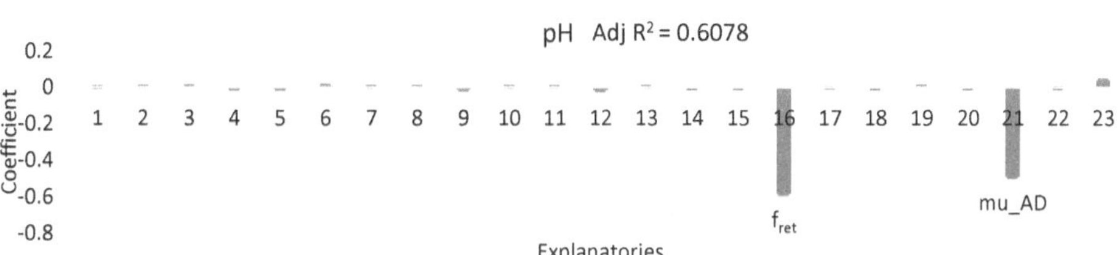

Figure D14: β coefficients for pH

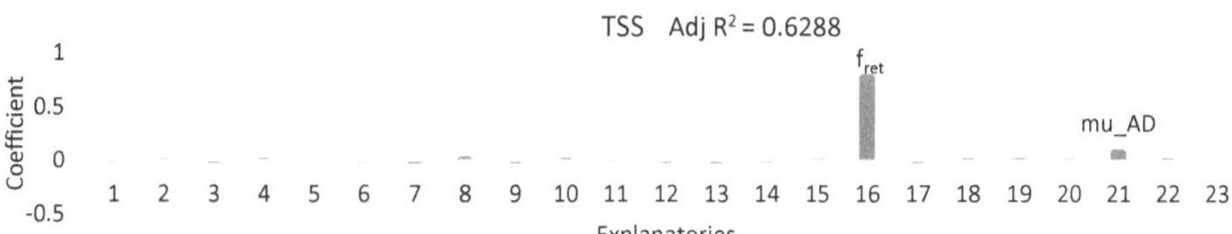

Figure D15:β coefficients for TSS

2. LASSO RESULTS FOR MODELS USING PRIMARY EXPLANATORIES WITH INTERACTION

The table below lists the explanatories or combination of explanatories (due to interaction) which are deemed most significant by lasso. Because only 3 variables, namely S_H, S_NH and pH, had an improved adjusted R^2 greater than 0.7 with the inclusion of interaction between primary explanatories, only these 3 variable results are shown here. It is evident that the most significant terms here are f_{ret}, mu_AD, mu_HM and i_N_Org_mol_perC.

Table D1: Lasso results for analysis with interaction terms

Response	Model	Lasso Penalty	Adj. R^2
S_H	-0.00298 + 0.667156(f_{ret}) + 0.426008(mu_AD) + 0.064038(mu_HM) - 0.269173(f_{ret}*mu_AD) + 0.063181 (mu_AD*mu_HM)	1.4	0.7366
S_NH	0.00722 - 0.64122(f_{ret})-0.041629(i_N_Org_mol_perC) - 0.487254(mu_AD) - 0.013316(f_{ret}*i_N_org_mol_perC) + 0.218617(f_{ret}*mu_AD)	5.6	0.689
pH	0.013449 - 0.577674(f_{ret})- 0.485371(mu_AD) + 0.31373(f_{ret}*mu_AD)	0.1	0.7123

3. FULL MODELS USING BOTH PRIMARY AND SECONDARY EXPLANATORIES, NO INTERACTION

The graphs below were obtained with models using all explanatories without any interaction terms. For each variable, two graphs are plotted. One graph plots all explanatories while the second graph only plots the primary explanatories. Because of the high β coefficients of the secondary explanatories, due to the high degree of association between the variable and the secondary explanatories, the graphs provide little information on the importance of the primary explanatories and hence the reason for the second graph under each variable. So, the second graph is the same information plotted as the first, but merely "zoomed in" on the primary explanatories to extrapolate their importance. Variables which attained a high

257

adjusted R^2 (greater than 0.7) without the inclusion of secondary explanatories were not reanalysed hereunder.

Figure D16: β values for Alkalinity showing primary and secondary explanatories

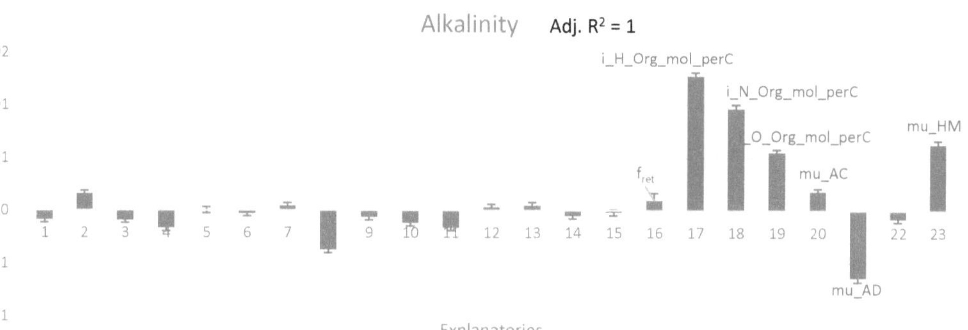

Figure D17: β values for Alkalinity showing primary explanatories only

Figure D 18: β values for S_CO3 showing primary and secondary explanatories

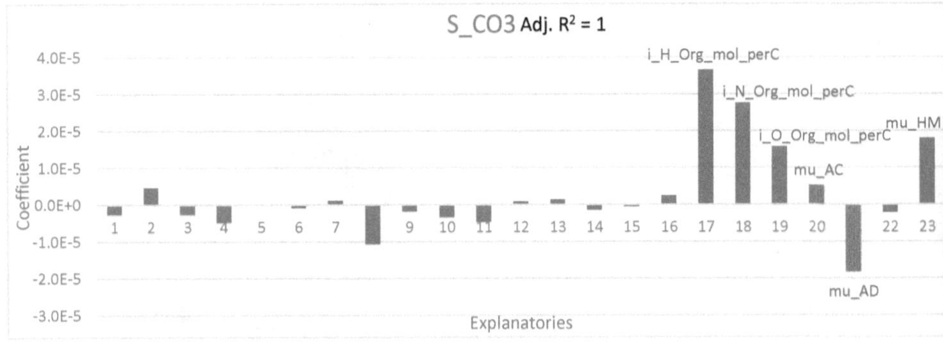

Figure D19: β values for S_CO3 showing primary explanatories only

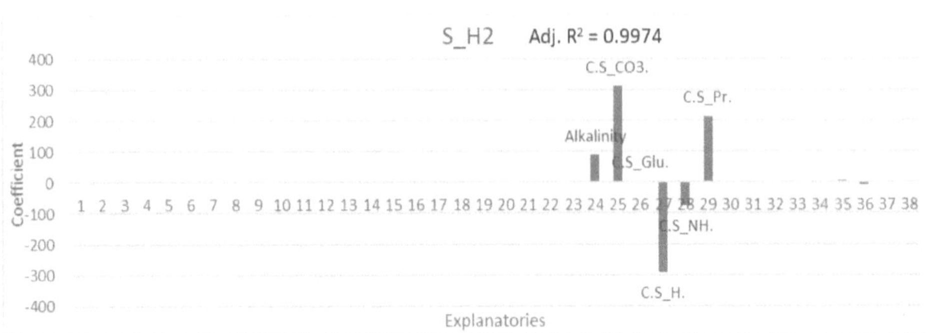

Figure D20: β values for S_H2 showing primary and secondary explanatories

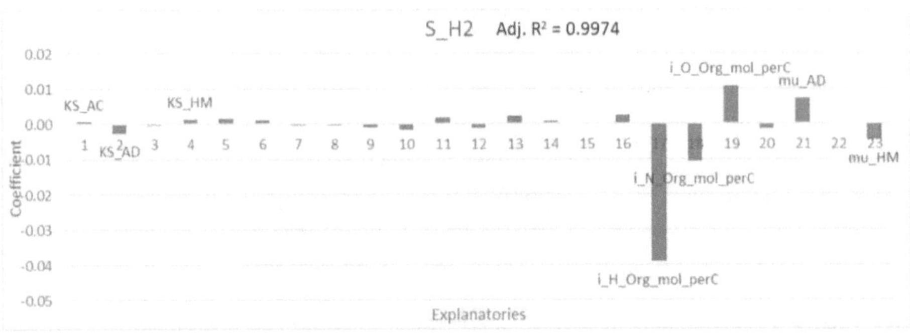

Figure D21: β values for S_H2 showing primary explanatories only

Figure D24: β values for S_Pr showing primary and secondary explanatories

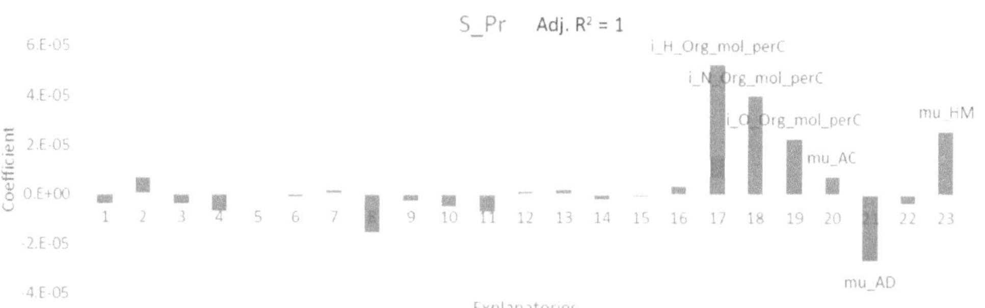

Figure D22: β values for S_Pr showing primary explanatories only

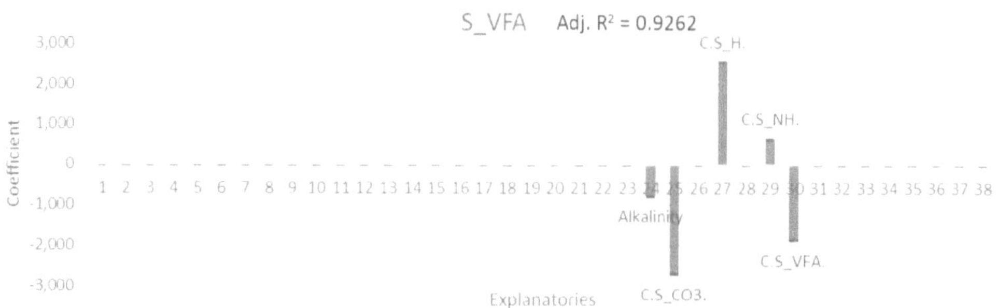

Figure D23: β values for S_VFA showing primary and secondary explanatories

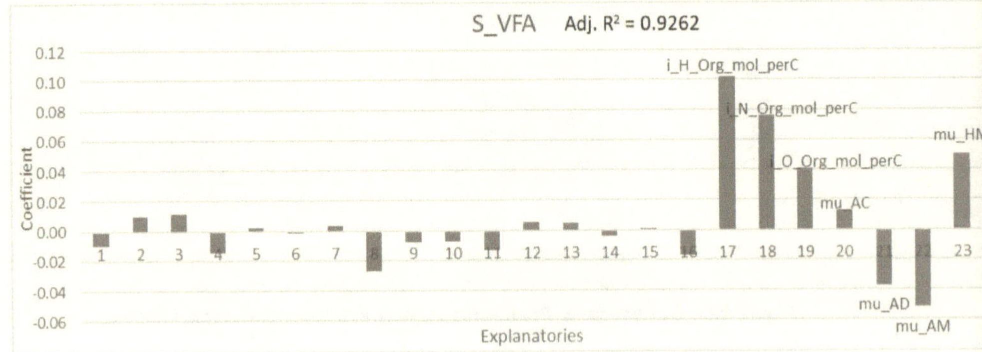

Figure D25: β values for S_VFA showing primary explanatories only

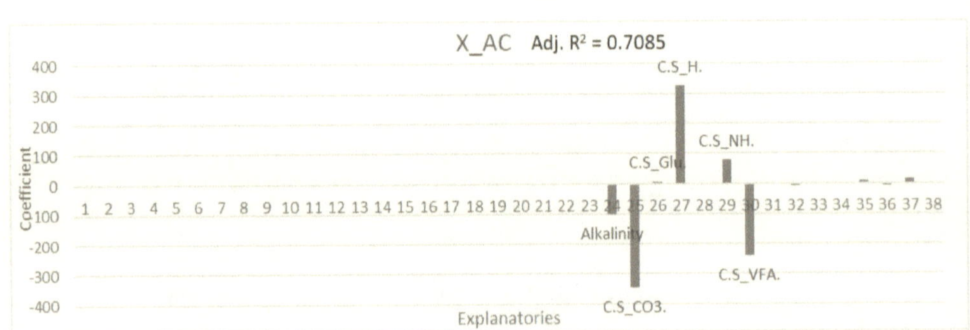

Figure D26: β values for X_AC showing primary and secondary explanatories

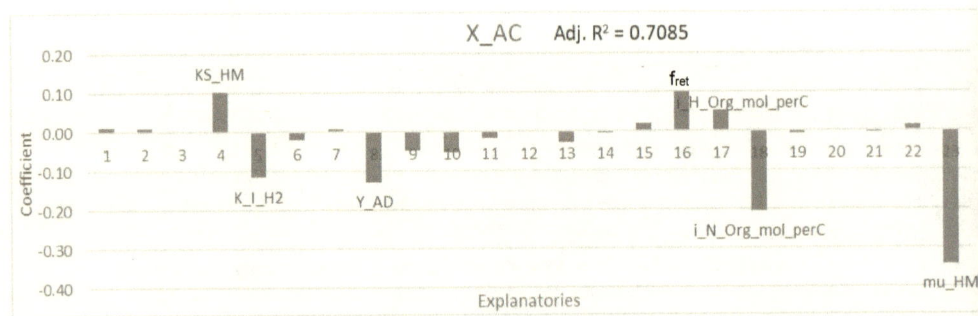

Figure D27: β values for X_AC showing primary explanatories only

Figure D29: β values for X_AD showing primary and secondary explanatories

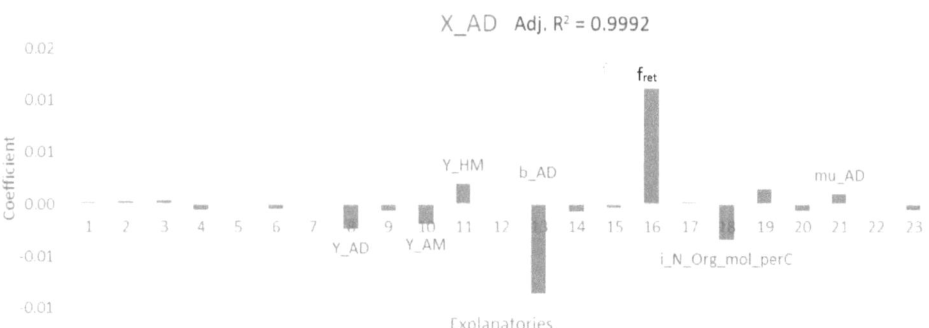

Figure D28: β values for X_AD showing primary explanatories only

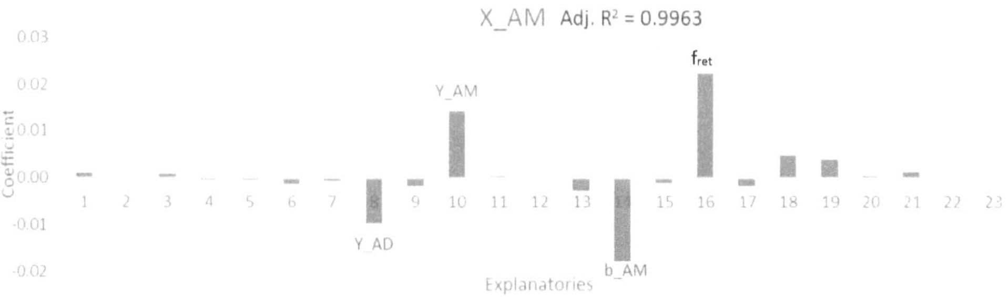

Figure D30: β values for X_AM showing primary explanatories only

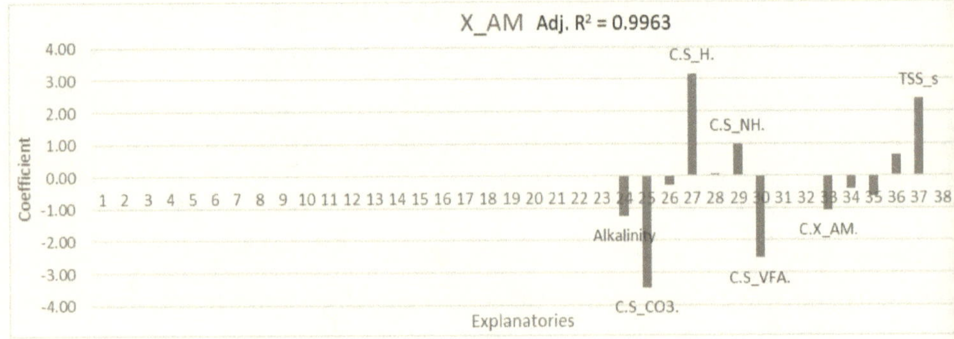

Figure D32: β values for X_AM showing primary and secondary explanatories

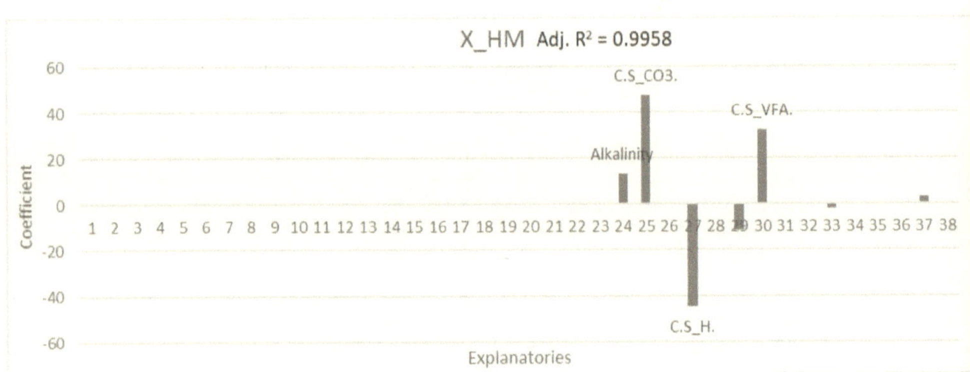

Figure D31: β values for X_HM showing primary and secondary explanatories

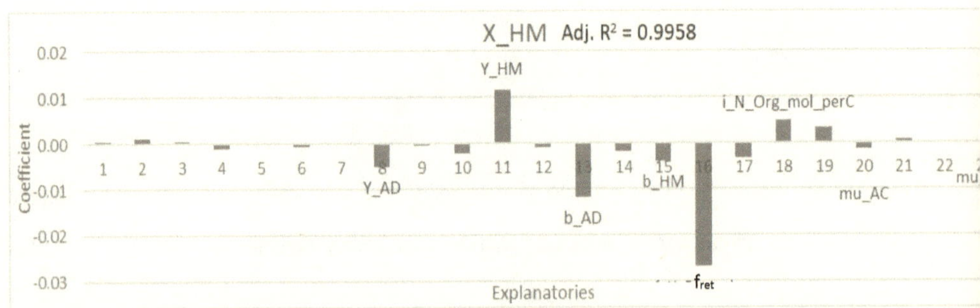

Figure D33: β values for X_HM showing primary explanatories only

4. LASSO RESULTS FOR MODELS USING PRIMARY EXPLANATORIES WITH INTERACTION

The results below document the reported explanatories based on lasso analysis. It should be remembered, as described above, that the secondary explanatories significantly hide the important primary explanatories due to the high degree of association between the variable in question and the secondary explanatories.

Table D 2: Lasso results for the full dataset (primary and secondary explanatories) for variables which did not obtain satisfactory results with simpler models

Output Variable	Explanatories	Lasso penalty	Adj R^2
Alkalinity	KS_AM, K_I_H2, Y_AD b_AC, b_AM, b_HM, i_N_Org_mol_perC, i_O_Org_mol_perC, mu_AC, mu_HM, S_CO3, S_Glu, S_H, S_H2, S_NH, S_Pr, X_AC, TSS, p_H	0.1	0.999
S_CO3	Y_AD, b_AC, b_AD, i_H_Org_mol_perC, i_N_Org_mol_perC, mu_AD, Alkalinity, S_Glu, S_H, S_NH, S_Pr, X_AC, X_AM, X_HM, Gasflow_CO2	0.1	1.00
S_H2	K_I_H2, b_AC, f_{ret}, mu_AC, mu_HM, S_H, S_Pr, X_AC, Q_GasflowG_CH4	0.1	0.85841
S_Pr	KS_AD, b_AC, f_{ret}, i_N_Org_mol_perC, mu_AD, mu_HM, Alkalinity, S_CO3, S_H, S_H2, S_NH, X_AC, X_HM, Q_GasflowG_CH4, TSS, p_H	0.45	1
S_VFA	KS_AC, KS_AD, KS_AM, KS_HM, K_I_H2, K_I_H_AD, Y_AC, Y_AD, Y_AH, Y_AM, Y_HM, b_AC, b_AD, b_AM, b_HM, f_{ret}, i_H_Org_mol_perC, i_N_Org_mol_perC, i_O_Org_mol_perC, mu_AC, mu_AD, mu_AM, mu_HM, Alkalinity, S_CO3, S_Glu, S_H, S_H2, S_NH, S_Pr, X_AC, X_AD, X_AM, X_HM, Gasflow_CH4, Gasflow_CO2, TSS, p_H	0.00	0.92616
X_AC	KS_AC, KS_AD, KS_HM, K_I_H2, K_I_H_AD, Y_AC, Y_AD, Y_AH, Y_AM, Y_HM, b_AC, b_AM, b_HM, f_{ret}, i_H_Org_mol_perC, i_N_Org_mol_perC, i_O_Org_mol_perC, mu_AC, mu_AM, mu_HM, S_Glu, S_H, S_H2, S_NH, S_Pr, S_VFA, X_AD, X_AM, X_HM, TSS	5.00	0.700
X_AD	Y_AD, b_AD, f_{ret}, Alkalinity, S_Glu, Gasflow_CO2, TSS	9.65	0.99790
X_AM	Y_AD, Y_AM, b_AM, f_{ret}, S_CO3, S_Pr, TSS	5.3	0.99468
X_HM	Y_HM, f_{ret}, Alkalinity, S_NH, TSS	5.00	0.98945
TSS	Y_AD ,Y_HM, b_AD, b_AM, i_O_Org_mol_perC, mu_HM, S_H2, S_NH, S_Pr, X_AC, X_AD, X_AM, X_HM	1.9	0.99967

5. MORRIS SCREENING RESULTS

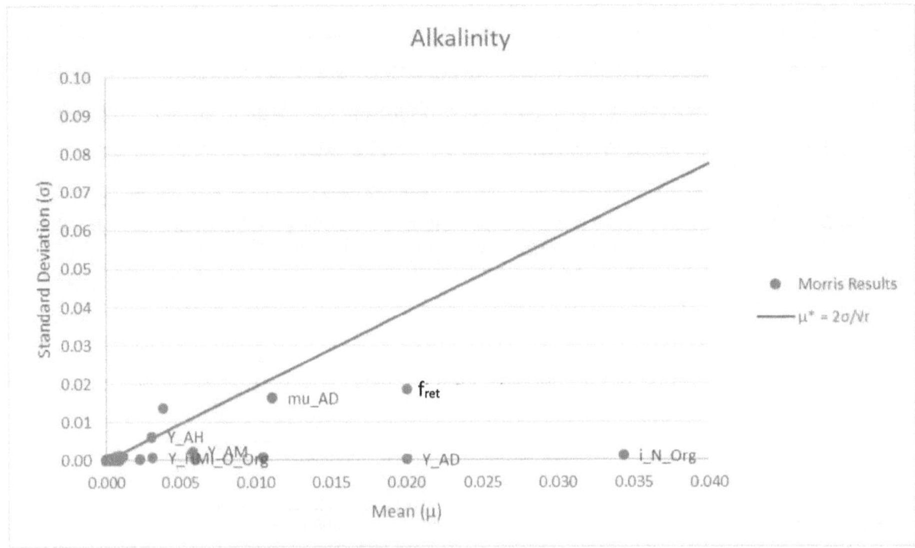

Figure D 35: Morris Result for Alkalinity

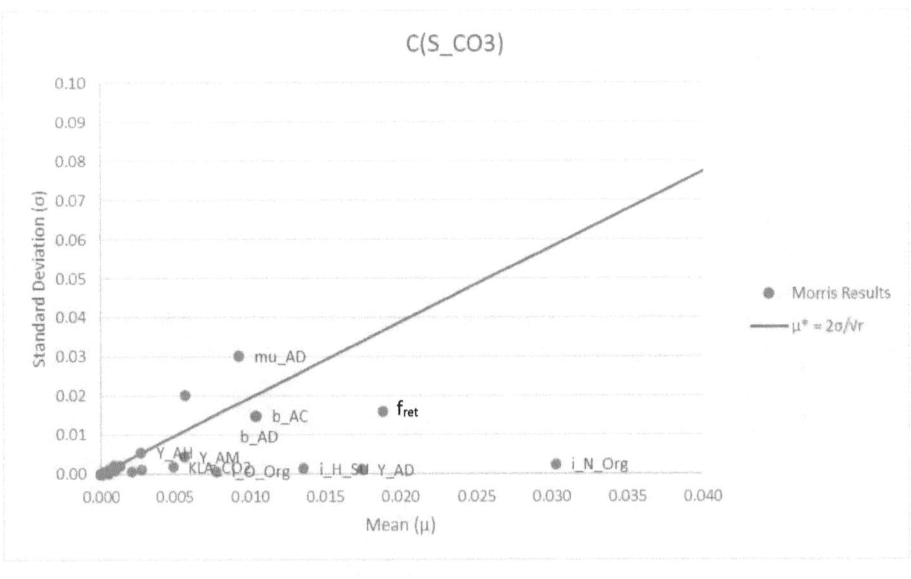

Figure D 34: Morris Result for S_CO3

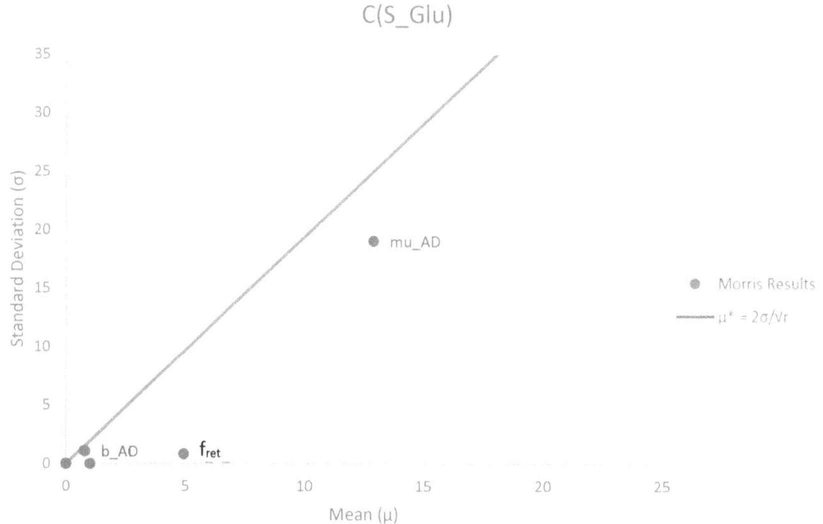

Figure D 36: Morris Result for S_Glu

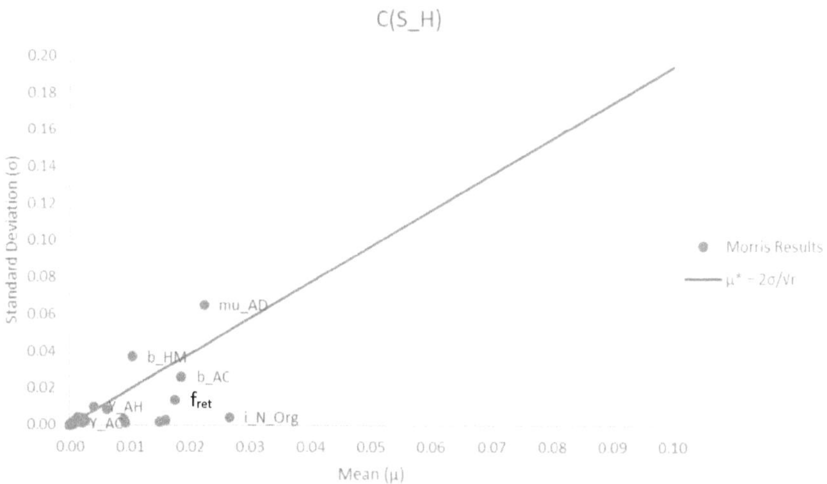

Figure D 37: Morris Result for S_H

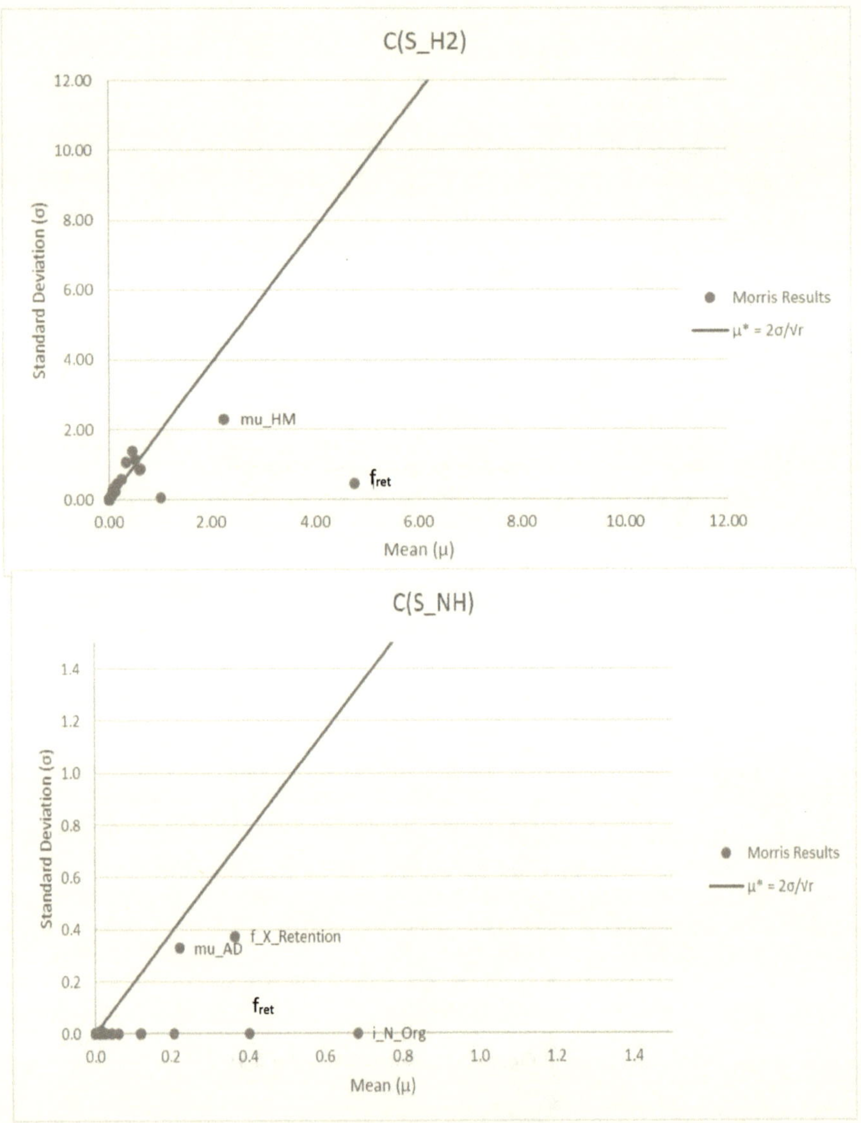

Figure D 39: Morris Result for S_NH

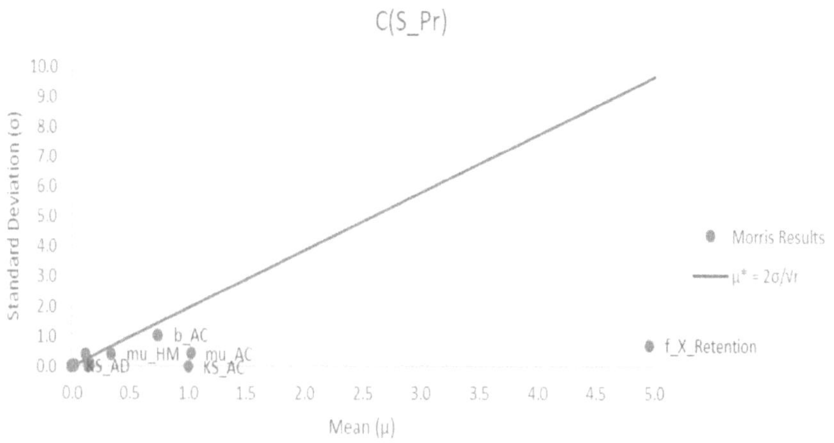

Figure D 41: Morris Result for S_Pr

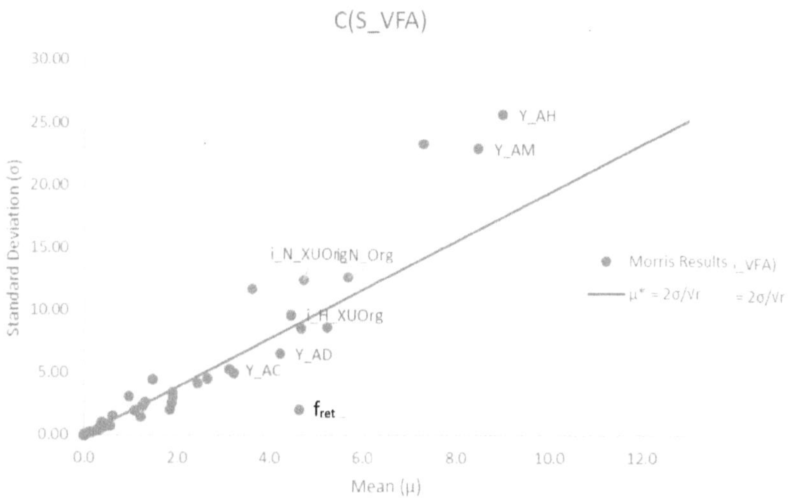

Figure D 40: Morris Result for S_VFA

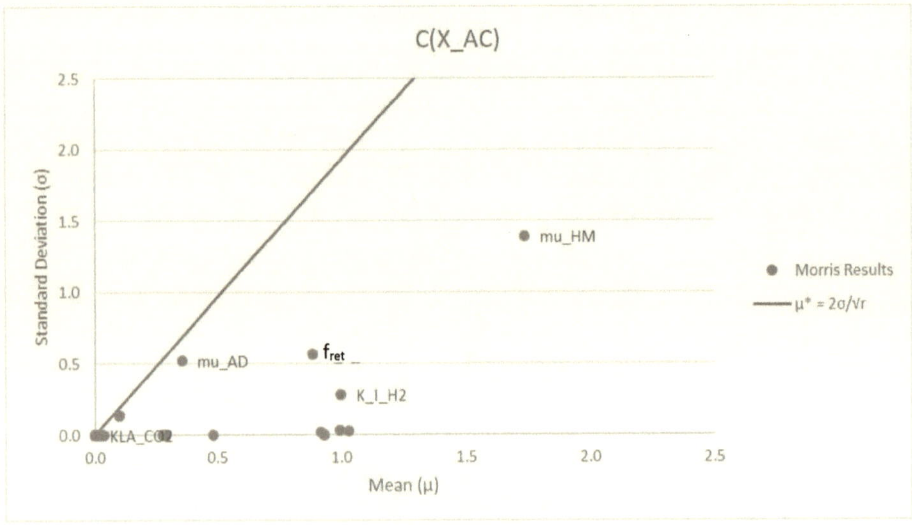

Figure D 43: Morris Result for X_AC

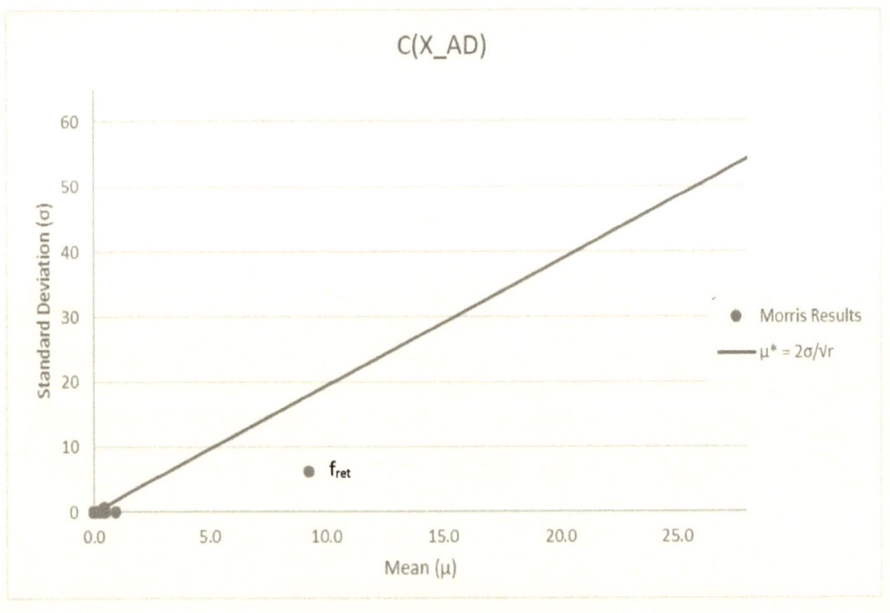

Figure D 42: Morris Result for X_AD

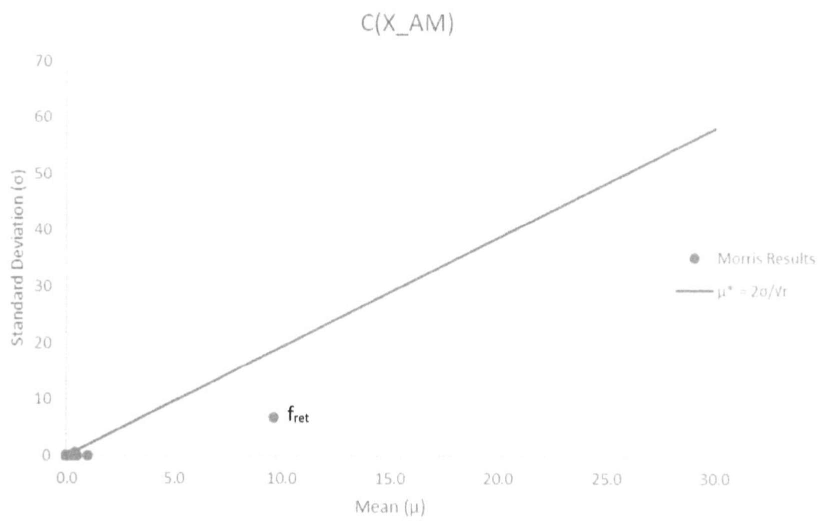

Figure D 45: Morris Result for X_AM

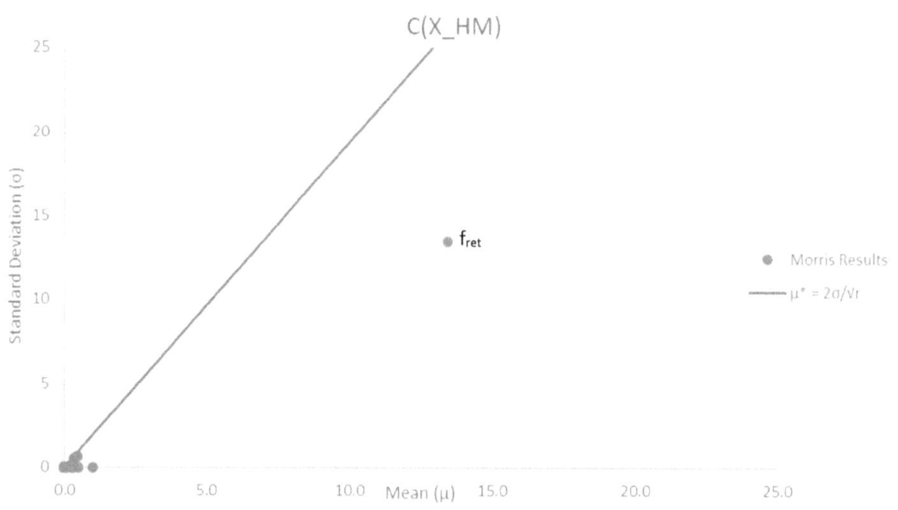

Figure D 44: Morris Result for X_HM

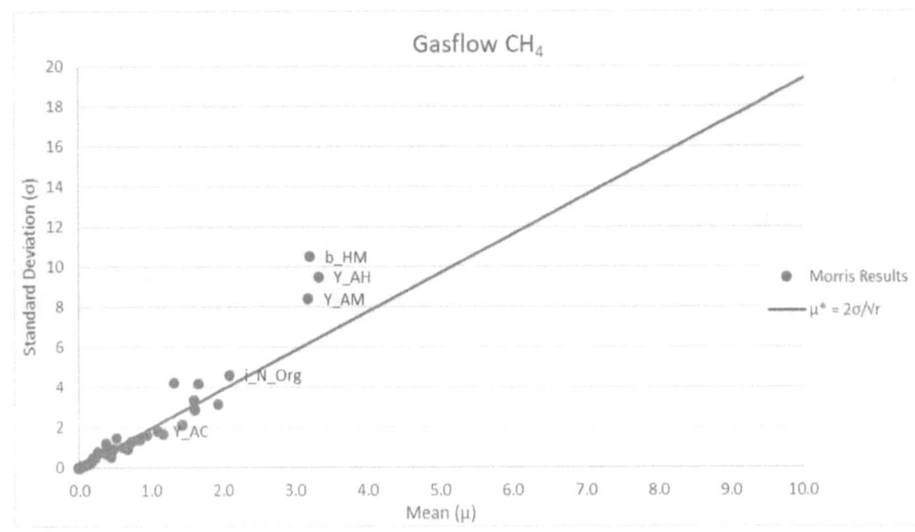

Figure D 46: Morris Result for Gasflow CH$_4$

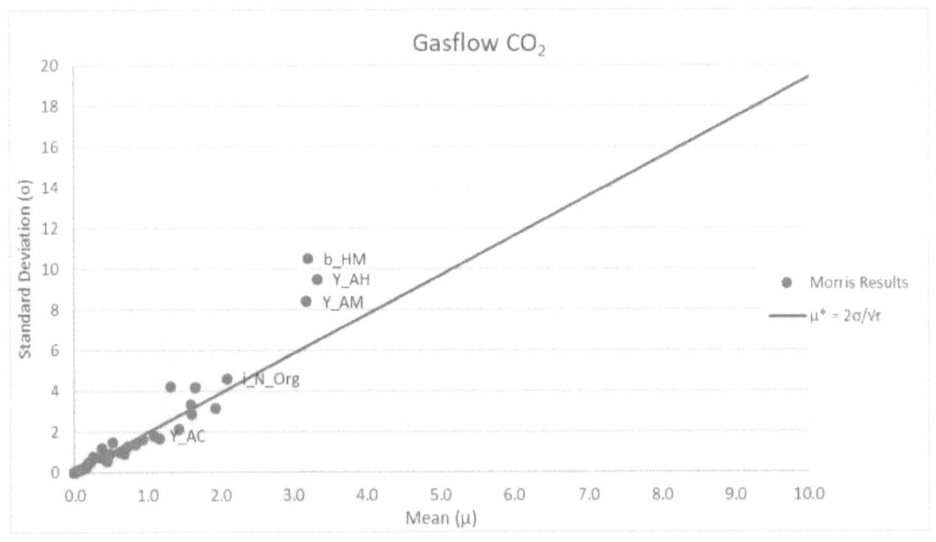

Figure D 47: Morris Result for Gasflow CO$_2$

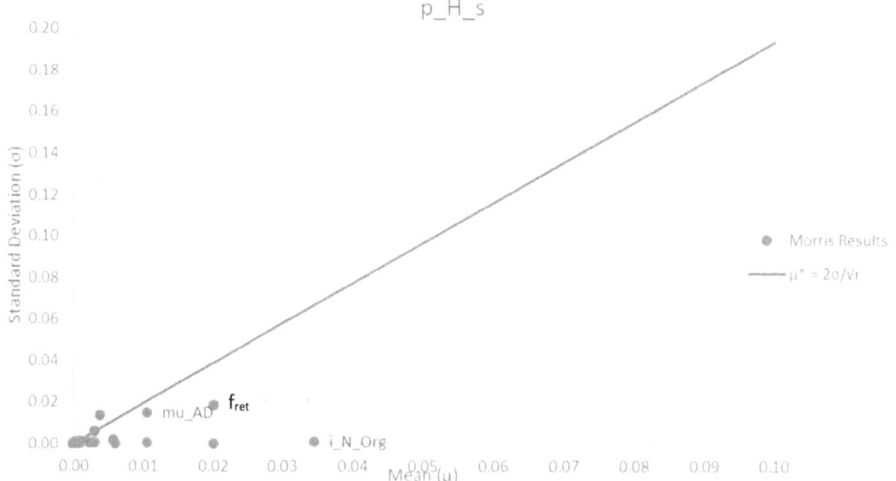

Figure D 48: Morris Result for p_H_s

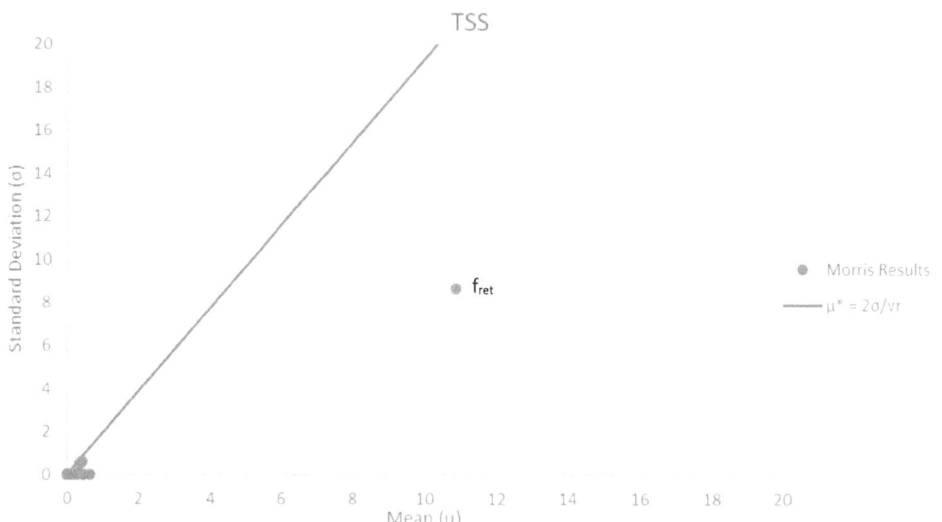

Figure D 49: Morris Result for TSS

www.ingramcontent.com/pod-product-compliance
Lightning Source LLC
LaVergne TN
LVHW042355190726
843493LV00005B/1024